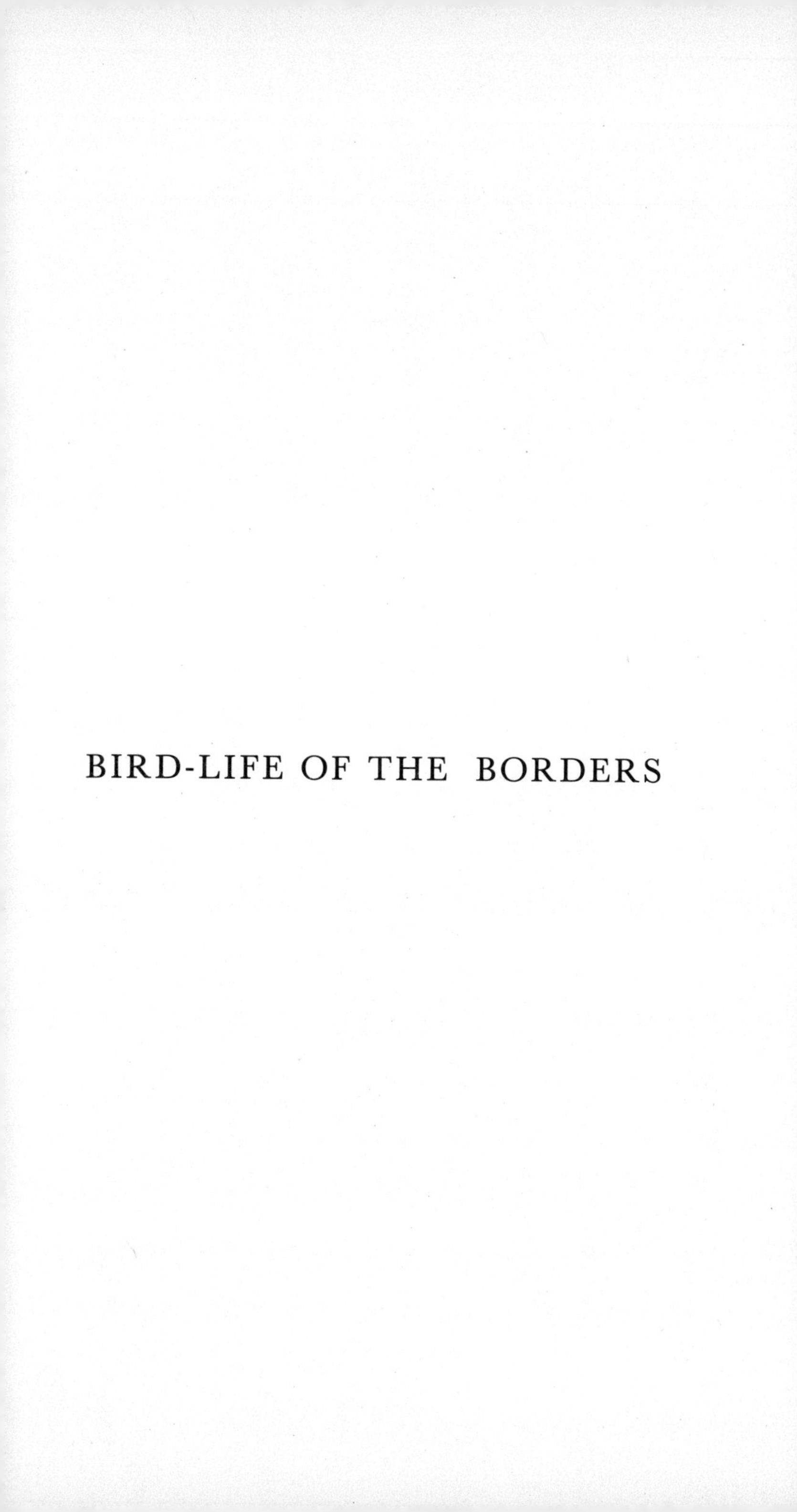

BIRD-LIFE OF THE BORDERS

BIRD-LIFE OF THE BORDERS

Records of
Wild Sport and Natural History
on Moorland and Sea

By
Abel Chapman

THE SPREDDEN PRESS
1990

First published 1889 by
Gurney and Jackson

This edition published 1990 by
The Spredden Press
Brockbushes Farm
Stocksfield
Northumberland NE45 7WB

Printed and bound by
SMITH SETTLE
Ilkley Road, Otley, West Yorkshire LS21 3JP

INTRODUCTION

Bird-Life of the Borders, Abel Chapman's first book, immediately brought him recognition as an author capable of communicating his knowledge and love of wilderness and wildlife to a wide audience, and his vivid and accurate descriptions of the ornithology of moorland, wetlands and shoreline of the Border country still have tremendous appeal. In his 'Memoir' of Chapman, his friend and fellow ornithologist George Bolam stated that '*Bird-Life of the Borders* ... is a classic muniment of observation and field work that, alone, would keep his name green through the ages'. Bolam could not have foreseen the current usage of the word green, yet both he and Chapman were precursors of today's environmentalist movement, providing detailed accounts of the fauna and flora of Northumberland and the Borders which provide a baseline for today's conservationists. Chapman had an acute awareness of the increasing threats to wildlife which were becoming apparent one hundred years ago, not only in Britain but also much further afield, and he regarded amongst his greatest achievements his role in the establishment of the Kruger National Park in South Africa, and in providing advice which saved the Spanish Ibex from extinction.

This concern for the environment provides us with a paradoxical view of the author of *Bird-Life of the Borders:* it is perhaps difficult, towards the end of the twentieth century, to reconcile the dual role of hunter-naturalist. Written when he was thirty-seven years old, *Bird-Life* communicates the enthusiasm for field sports which he had pursued for twenty years or more, yet indicates a deep understanding of the biology of birds and their habitats. Abel Chapman followed a long line of sportsmen who were accomplished naturalists: his maternal grandfather, Joseph Crawhall, was an accurate shot on the grouse moors of Hexhamshire, and a founder member of the Natural History Society of Northumberland, Durham and Newcastle upon Tyne in 1829. His uncle, George Crawhall, was described by Chapman as 'a typical sportsman of the old school - the mentor to whom I owe the best of grounding in field-craft'. Although in his preface he is self-effacing about his skill as a hunter and fisherman, the collections of trophies and specimens he made speak for themselves. Chapman must be seen as a man of his time, when attitudes to hunting were long established and no conflict was recognized between the pursuit of game-birds or wildfowl and the study of natural history. At this time, when collection-building for museums was in full swing, naturalists were probably having a greater effect on the bird fauna than the sportsmen. The famous Newcastle ornithologist and taxidermist, John Hancock, used a gun with considerable skill to build up his knowledge of the distribution of birds and add to his collections. Chapman makes reference to the greed of collectors in reducing the local raven population to critical levels, and the dramatic impact of gamekeepers on birds of prey, such as the hen-harrier and peregrine, which he regarded as unnecessary and regrettable.

The pages of *Bird-Life of the Borders* are enlivened by Chapman's sketches, lending much charm to the book. However, the author is somewhat disparaging about his skill as an artist, writing of the illustrations that the 'rough pen and ink drawings ... reproduced by photo-zincography ... are intended to serve as character-sketches rather than as portraits, and have no pretensions either to scientific accuracy or artisitic merit'. There can be no doubt that Chapman's drawings trap the moment for us, the birds caught in lifelike postures, the result of years of meticulous observation and recording. He used rapid field sketches made in his notebook as the source of the illustrations which appear in his published works. Some of these original notebooks are held in the archive collections in the Hancock Museum, and even the earliest of them, dated 1869, indicate his ability to capture places, people and wildlife with a pencil. His acid wit, his interest in anecdote and incident, as well as his rapid sketches, pepper the notebooks which record his travels from the Arctic to Africa throughout a life of adventure.

Born at Silksworth Hall, Sunderland, on 4 October 1851, Abel was the eldest of the six sons and two daughters of Thomas Edward Chapman and his wife Jane. From 1865 to 1869 he attended Rugby School, where he acquired his knowledge of the classics, and the friendship of F. C. Selous, whose inspiration generated Chapman's love of travel and adventure. Later, in 1914, he collaborated with Selous and J. G. Millais to write *The Big Game of Africa and Europe*.

Returning to northern England, Chapman entered the family firm, the Lambton Brewery, Sunderland, and became much involved in the wine trade, travelling to Portugal, Spain, Morocco and Southern France on business. He made good use of these trips to broaden his knowledge of wildlife, indulge in shooting and fishing, and to add to his growing collections: throughout his long commercial career (he retired when the Brewery was sold to J. W. Cameron in 1897) he appears to have spent most of his leisure time observing nature. The result of all his collecting efforts have enriched the Natural History Museum in London, where important reference specimens — a new species and representatives of distinct sub-species and geographical races — have been lodged, and the Hancock Museum in Newcastle, for which Chapman had great affection. He was present at the opening of the building in 1884, and joined the Natural History Society of Northumberland, Durham and Newcastle upon Tyne that year, becoming a staunch supporter, frequently making considerable financial contributions to the Society. His trophy collection (now displayed as 'Abel's Ark'), and other miscellaneous specimens were bequeathed by him to the Museum in 1929; his diaries and notebooks were acquired later.

During his annual visits to Spain, Chapman became a close friend of Walter J. Buck of Jerez, the two men sharing a passion for wildlife. In 1882 they became joint lessees of the Coto Doñana, a forty-mile long stretch of coast north of the delta of the river Guadalquivir. Chapman discovered the breeding grounds of the flamingo in the nearby marismas, the first ornithologist to do so according to his biographer, T. R. Goddard. Many of their experiences of hunting and observing Spanish wildlife were described by Chapman and Buck in *Wild Spain* (1893) and *Unexplored Spain* (1910). The Coto Doñana, was acquired by the Spanish Government in 1963 and 1965; some 6,500 hectares are now managed as a national nature reserve.

In August 1881 Abel Chapman made his first journey north, to Spitzbergen. He became fascinated by Scandinavia, and in seventeen years made no less than twenty-three expeditions to the far north, mainly to Norway, but also to Sweden and Denmark. His brother, Alfred Crawhall Chapman, accompanied him on many of these journeys. Alfred was also a dedicated diarist and illustrator, and the brothers' detailed accounts of these adventures found in their notebooks (now in the

Hancock Museum) were compiled in the book *Wild Norway,* published under Abel's name in 1897.

In 1899, with his brother Walter, Abel Chapman made his first visit to Africa to shoot big game. This visit — to South Africa — was cut short by the Boer War, the brothers having a difficult homeward journey via Mozambique and Madagascar, working their passage in a Red Sea trading boat. Undaunted, Chapman returned to British East Africa and Equatoria (now Kenya and Uganda) in the summer and autumn of 1904 and in the spring of 1905. His experiences were vividly recounted in his book *On Safari* (1908). His attention then focused on the vast unexplored African country of Sudan, which he visited in the winters of 1912-13 and 1913-14; he returned after the first World War in 1919. *Savage Sudan* (1912) is full of incident, anecdote and adventure, embellished with Chapman's sketches, an account which highlights his love and understanding of wild and remote places and their people. Chapman's 'African period' are also featured in his later books: *The Borders and Beyond: Arctic, Cheviot, Tropic* (1924) and *Retrospect: Records and Impressions of a Hunter-Naturalist* (1928).

In Chapman's day the use of the punt-gun was regarded as an acceptable means of shooting ducks and geese, and in *Bird-Life* he describes some of the hardships of a bitterly cold night spent lying in the bottom of a boat waiting for the moon, and the ducks, to appear. His classic book about punt-gunning, *The Art of Wildfowling,* was published in 1896, but an indication of Chapman's total fascination with wildfowl and wildfowling can be found in *Bird-Life.*

Altogether, including the second edition of *Bird-Life of the Borders* (1907), Chapman wrote ten major works. T. R. Goddard, in his detailed account of Chapman's life, comments that 'they were not compilations, but faithful and graphic records of his own personal experiences, representing a vast amount of time and labour in the field'. Another book, *Memories of Four Score Years Less Two,* was in progress in 1929, and was prepared for publication by his friend George Bolam and the artist W. H. Riddell.

Despite his extensive foreign travels, Abel Chapman had a special regard for Northumberland and the Borders, making his home in the region at Houxty on the North Tyne in 1898.

According to Bolam, Houxty, with its plantations, moorland and gardens, became a second Selborne, attracting birds, animals and naturalists in profusion. With blackcock numerous on the estate, and salmon in the river, it is little wonder that Chapman had great affection for it; his manuscript *Fauna Houxtiensis* records one-hundred-and-thirty-four species of birds seen there, one-hundred-and-ten of them identified from the window! Chapman died at Houxty on 23 January 1929, and was buried in the churchyard at the little village of Wark.

In *Bird-Life,* Chapman defines the geographical boundaries of his Border Country, 'whose peculiar beauty is rather characteristic than sensational', as from the Cheviot to the Solway. In *The Borders and Beyond* he states 'The Borders were my first love and today, sixty years later, remain my last. Never, during that long period, has the charm of the Cheviots and of Ettrick Forest, with the far-flung mountain-land that lies between, abated or suffered eclipse'. There can be no doubt, however, that he saw many changes in the landscape of the Borders during his long association with it, changes which affected the wildlife of the region. In the hundred years since the publication of Chapman's first book, the pressure on the land has increased dramatically; moorland has been extensively claimed for forestry, the waters of his North Tyne have been dammed for the Kielder Reservoir, flooding some of the most attractive riverside habitat in Northumberland, whilst lower-lying land has been subjected to agricultural improvement by draining, re-seeding and the use of fertilisers. The pattern of rough hill grazing and heather moor which Chapman loved has been much altered and reduced. It is not just a history of the loss of habitat, but of increasing human access: better communications and the demand for outdoor recreation have destroyed the wild and remote nature of the hill country. Changes on the coast have been primarily associated with recreation and tourism, again reducing the wilderness qualities which Chapman so obviously relished. The changes to the marine ecosystem caused by overfishing and pollution are less visible, but of considerable concern for the long-term survival of coastal species, especially the wildfowl and waders which Chapman describes.

In spite of the changes in land use and increasing pressures on the environment, Chapman's descriptions of the changing pattern of the ornithology of the Borders through the year remain largely accurate. The status of certain species has, however, altered dramatically. The blackcock *Tetrao tetrix* was a bird for which Abel Chapman had considerable affection, indeed one of the major attractions of Houxty on his first visit with his legal adviser was this species: 'See those birds? ... they are blackcock, and we'll buy the place!' Once relatively common in the uplands of England and Wales, the populations resident in the area immediately south and east of the River South Tyne and to the north of the Yorkshire Dales constitute the one stronghold which is substantial enough to have any realistic prospect of long-term survival. Blackgame now seem sparse in the areas where Chapman knew them best: in recent surveys few leks were located in the North Tyne, in the Kielder area and in North Northumberland. The reasons for this decline are complex, but changes in land use are certainly implicated, the species requiring a mosaic of habitats, including rough grassland, heather moor, bog and open woodland, a pattern which has become impoverished over the years.

Chapman's observations on some species remain totally accurate. For example, he notes that the sudden, irregular incursions of some species which are thought of as scarce are a regular feature of ornithology, and uses as an example the grasshopper warbler *Locustella naevia* which he first heard at Silksworth Hall on 3 May 1882. This species is still regarded as an uncommon passage visitor and breeding species, most likely to be heard in late April and early May. Chapman suggests that the pied flycatcher *Ficedula hypoleuca* 'should probably be considered rather as a local species than a very rare one in the north of England'. He first saw the species — again at Silksworth — in 1855, and it was his opinion that 'this handsome little bird appears to be increasing in numbers'. He was correct: the pied flycatcher is now described as a well-represented passage visitor and breeding species, widely distributed in deciduous woodland.

Reference has been made to Chapman's regret at the persecution and over-exploitation of certain bird species, particularly birds of prey, and other species then regarded as pest species or competitors with man such as the raven. Probably the most significant change since the publication of *Bird-Life of the Borders* has been the changing attitude to predatory species, accelerated by the concern for their long-term future during the pesticide scares of the 1960s. Public awareness, and the increasing pressure exerted by specialist agencies such as the Royal Society for the Protection of Birds, has led to a change which Chapman could never have foreseen. He paints a black picture for the peregrine, which he regarded as a very rare bird in Northumberland, and for other birds of prey:

> The Wild moory hills and rugged crags which Nature assigned for its dominion, and where in years gone by this fine Falcon, together with the Buzzard, the Raven, and the Harrier, regularly nested, will in all probability soon know it no more. Rightly or wrongly, man has usurped the functions of Nature in adjusting the balance of life ... it does appear regrettable that the fate of the few survivors ... should ... be left at the mercy of ignorance and prejudice.

Chapman could cite only the merlin as being 'tolerably abundant'. One hundred years later, in spite of many years of persecution, the advent of organochlorine pesticides, increasing disturbance and continued pressure from collectors, the Borders are still a home to birds of prey. However, there are no grounds for complacency. Peregrine, merlin, buzzard and hen harrier are still under threat, and remain rare breeding species.

Abel Chapman, in all his books, but no more so than in *Bird-Life of the Borders,* conveys an enthusiasm and love of nature. He lived at a time when attitudes to wildlife were slowly beginning to change from exploitation to understanding and protection, but also at a time of increased demands on the land and its resources. By communicating his feelings of wonder, excitement and fascination for wilderness, and expressing his concerns for its survival, Chapman demonstrates that behind the tough, hunter-naturalist facade was a sound conservation philosophy.

PETER DAVIS
The Hancock Museum
The University
Newcastle upon Tyne
April 1990

We would like to thank the Natural History Society of Northumberland very much for allowing us to reprint from their first edition of *Bird-Life of the Borders.*

PREFACE.

THE substance of the following chapters is based on observations extending over a series of years, and accumulated while serving a long apprenticeship to rod, fowling-piece, and stanchion-gun, by "fell and flood," in the Borderland.

The use of the plural pronoun is intended to include my uncle, Mr. G. E. Crawhall, and my brothers, with whom many of my sporting days have been spent. While freely expressing opinions on sport and cognate subjects, I distinctly disclaim any pretensions to special skill therein, beyond the average: nor are the few narratives of sporting incidents introduced with a view to "blowing my own trumpet," but as the best means of illustrating certain phases of bird-life, and of the practice of wildfowling afloat. Like a finger-post, one may point to a goal which one is never permitted to attain.

A few of the chapters have appeared (substantially) as articles in *The Field*, others in the *Pall Mall Gazette*, and to the editors of those journals I am indebted for the privilege of reproducing them. To Mr. Howard Saunders, for his extreme good-nature in revising the proof sheets, and invaluable advice thereon, I owe no small debt of gratitude.

The illustrations—rough pen-and-ink drawings by the author, reproduced by photo-zincography—are intended to serve as character-sketches rather than as portraits, and have no pretensions either to scientific accuracy or artistic merit.

What I venture to claim for this book is that, so far as it goes, it traces, throughout the year, the life-histories of many of the most valuable and interesting birds—and, especially, of those wilder forms whose remote and desolate haunts, and intensely wary natures, have ever opposed the utmost difficulties to their study and close observation. I have endeavoured to treat the subject from a broad, rather than a local, standpoint. Ornithology loses half its charm when restricted by artificial boundaries—the feathered world recognize none but those of natural or physical adaptibility. An undue development of the prevalent system of "county ornithology," with its arbitrary limits and restricted "faunal areas," must have, in my opinion, a tendency to stunt the true interest of this study as a whole, and to narrow its scope.

Birds are cosmopolitans: those which one day we regard as British, may be African or Asiatic the next; the true home of very many lies amidst Arctic solitudes; and for many more our islands only provide a temporary refuge from the intolerable summer-heats of the tropics. Hence a full appreciation of ornithological science involves considerable foreign research—or better, foreign travel, far more than has fallen to the lot of the author. Every opportunity, however, has been so utilized, and having some half-dozen times crossed the North Sea in different directions, once the Arctic Ocean, and thrice the Bay of Biscay, I have explored a few of the wilder and more remote regions of Europe from Southern Spain to Spitzbergen. Of these foreign experiences, it is my hope some day (should the present venture succeed) to write some account, and very probably they may prove the more interesting and eventful. Perhaps I have undertaken the hardest task first; but for the present I beg to commend this volume to the kindly indulgence of the reader, in the confidence that it embodies the results of much careful work and thought.

ROKER, SUNDERLAND,
December 17, 1888.

CONTENTS.

LIST OF ILLUSTRATIONS.

BIRD-LIFE OF THE BORDERS.

SPRING-TIME ON THE MOORS.

Without going beyond the boundaries of our island, there yet remain wild corners which are neglected and all but unknown: the beauties of many a spot whose charm and intrinsic merits are as deserving of attention as those for which the tourist crosses the seas and seeks the uttermost parts of the earth remain ignored. Of these the Borderland is one. Stretching from Cheviot to the Solway, the wild uplands of the Border cover a broad area in either country, and include, here and there, scenery which it would probably be difficult to match within the four seas, though their peculiar beauty is rather characteristic than sensational, unique rather than "clamant," if I may borrow Professor Geikie's expressive term.

The area covered by these observations I would define as the mountain-region which remains unaltered by the hand of man—the land "in God's own holding"—bounded by the line where the shepherd's crook supplants the plough; and heather and bracken, whinstone and black-faced sheep replace corn, cattle, and cultivation; where the Pheasant gives way to the Grouse, and the Ring-Ouzel dispossesses the Blackbird: the region of peat, as distinguished from soil, of flowe, moss, and crag, of tumbling burn and lonely moorland, clad in all the pristine beauty of creation.

My whole area, in short, is one great sheep-walk, where grouse and sheep outnumber man in the proportion of hundreds, or thousands, to one. On the higher fell-ranges it takes two acres to support each sheep—there are barrens where even this proportion is largely exceeded—and the minimum for a man may thus be roughly set down at five hundred to a thousand. Thus the hill-country is all but uninhabited, abandoned to shepherds and flock-masters, where sequestered homes lie scattered among the recesses of the hills. A hardy race are these to whom *ovis bidens* is the *præterea nihil* of life, for the more severe the weather the greater the necessity to "keep the hill": and kindly and hospitable they are forbye, as the belated traveller can testify who has lost his way among the mists of cloudland on the fell-tops.

An initial difficulty in describing the bird-life of any given area throughout the year, is to decide at which point to begin. New Year's Day suits human purposes well enough; but Nature provides no break in her cycle, and no single point of time can be found at which her various operations can start level. Hence these chapters will necessarily partake something of the character of those golden serpents which one sees made into ladies' bracelets, and which complete the continuity of their circle by taking a large piece of their tails into their jaws.

The opening months of the year are uninteresting and uneventful on the moors. There is but little perceptible change from the conditions which prevailed during November and December, and an outline of the ornithological features of those months will be found in due course. Hence there is little attraction to detain us till the advent of spring, or of the vernal influence, at which somewhat indefinite period these notes will therefore commence.

Springtide is a subject on which, from time immemorial, poets and those of vivid imagination have delighted to descant. And, truly, there is a charm in the idea of the rejuvenescence of all Nature's productions at this season—when everything becomes revivified, and new life springs afresh in bird, beast, and plant—which is generative of poetic instinct. In these chapters, however, the author is necessarily restrained from indulgence in any sentimental effusiveness beyond what may be dictated by the logic of facts. For in the north of England—and especially on its moory uplands—the term SPRING represents rather a chronological definition than the embodiment of an idea calculated to inspire, from the character of the period it defines, any very high-flown sentiment of poesy. The months of March, April, and often May, include (with the possible exception of September) the worst and most unpleasant period of the year, as regards weather, on the northern hills. Up to the very end of May we are still liable to snow-falls, and the high lands often lie as white then as in December. If one of these months chances to be bright and fine, the others do extra penance to the Nimbi for its errors, and one has to be thankful for single mercies. Jupiter Pluvius holds sway, and as day after day, and week after week, one looks out at the cold, north-easterly sleet driving along the hill-sides, and the pitiless, pelting *mistraille* shrouding their summits from view, and sending down the burns in top-flood, there is little, it will be admitted, to provoke sentimental outbursts of enthusiasm at the new-born glories of the "glad season," or the revivifying effects produced by the increasing powers of warmth and light.

And yet, unkindly as may be the elements, but little, if any, difference is produced by them on the seasonal progress of Nature's economy. Thus the Raven goes to nest at the end of February, utterly careless of the temperature—of the thermometer standing several degrees below the freezing point, and of a foot of snow enveloping the hills. She knows her appointed time, and cares for none of these things. And so it is with most of Nature's creatures. The sequence of events, each at its appointed season, goes on with marvellous regularity and with imperceptible regard for extraneous conditions. It is true the grass on the northern hills hardly commences to grow before June—the curved head of the bracken only emerges from the peat during that month—and the heather shows but little change from the black and lifeless hue it assumed in October till towards the period which, by the almanac, should be called summer. For these, spring is simply non-existent; but with the higher forms of life it is different. The moor-birds regard only their fixed and appointed seasons. They arrive, pair, nest, and hatch their young almost without reference to climatic conditions, and many a moorland chick first sees the light in an atmosphere and under circumstances which would appear necessarily fatal to its tender life. Small wonder that the vast majority of the strong-winged birds—the ducks, geese, and other wildfowl—should prefer the Arctic regions for their *incunabula*. The words may cause a shudder to those who only associate them with thick-ribbed ice, with intense cold, and manifold forms of death. Yet, though the Polar summer is short, it is unquestionable that the Arctic lands, with their three months of uninterrupted sunshine, their boundless wastes of marsh and moor, and profuse wealth of plant and insect-life, afford conditions of life to the feathered tribes infinitely more favourable than does the spring climate of our temperate zone.

But I must not do injustice to the season, and would be drawing too gloomy a picture of the North British springtime if I omitted to mention the few spells of bright and warm days which, at uncertain intervals, do occur to break the monotony of even the most inclement springs. Oases in a desert they may be in many cases; but not for that reason is their advent the less welcome and delightful—quite the reverse. I am not alluding to those shams, those deceptive spring-like days when brilliant sunshine co-exists with a biting north-easter; when April showers descend in the form of fine snow or cutting hailstones; when one is baked in the shelter and frozen in the shade—such days are as false and illusory as they are common at this season. They are, perhaps, preferable to the fog and rain, but bear no comparison whatever to the truly vernal hours when the winds blow soft and warm from the west and south with the first touch of the zephyr in their breath.

On such mornings as these, when the sunshine bathes the water-logged moors in unwonted warmth, drying the dripping heather and moss, every creature appears inspired by the spirit of the vernal season. The moor-birds pipe and whistle in a wholly different key to their querulous notes of yesterday; visibly they luxuriate in the genial change. Under the cold and humid conditions of atmosphere which have hitherto prevailed, one can hardly form any very close acquaintanceship with them. One only hears their wild alarm notes as they spring, unseen in the fog, far away. Now, under the genial influence of warmth and a dry atmosphere, they cease to resent man's intrusion on their domains, and go about their domestic duties almost regardless of his presence, though close at hand. The wilder spirits—those irreconcilables that are impregnated, as it were, to the very marrow with inherent fear and suspicion of our race—such as the Mallard and the Curlew, may still think it necessary to keep a gunshot or two away from the intruder; but even these seem to do so half-unconsciously—merely from force of habit and association, and not at all in an *offensive* manner. The game-birds, the Plovers and the Teal, now abandon nearly all their hybernal shyness, and tacitly recognize the temporary suspension of hostilities. The trout, also, in the hill-burns which have hitherto disregarded all the attractions of insect-food—real or artificial—grubbing about on the bottom for their livelihood, now roll and play on the surface, in the glancing waters and in the heads of the streams. Everything, in short, man included, feels the exhilarating influence of the day, and enjoys it all the more from the knowledge that it may be very transient. Nothing, indeed, is more delightful than the rare spells of fine warm weather which do occur at this season—early spring—when the winter appears at last to have passed away, and the atmosphere becomes resonant with a chorus of wild bird-notes, and redolent with the fragrance of the heather-burning.

Where development depends on so extremely variable a factor as our spring climate, its course is necessarily very irregular. Up to the end of March there is but little conspicuous change from the bleak and wintry aspect of the moors. Many of the spring birds are there, it is true, but at first they are restless and shy. The spring element of trustfulness and confidence has not yet appeared, and the whirring forms of the grouse are still seen spinning away as wild as in November. Indeed, it is not till May that the true spirit of the vernal season is fully developed. In mid-April the only signs of vegetation are the catkins on the willows and osiers. At the end of that month the hardy birch and alder may show some symptoms of returning greenery; but the heather remains as black and as cold as ever, and the grass, rushes, and fern are but the dead and withered remnants of last year's growth, colourless and blanched by the weather, and flattened down by the weight of the winter's snows.

ARRIVAL OF THE SPRING-MIGRANTS,

WITH SOME OBSERVATIONS ON MIGRATION.

AMONG the earlier signs of returning spring, is the commencement of migration—a phenomenon so complex, and yet so deeply interesting, that I propose making a few remarks on its scope and on the causes which produce it, even though it may be at the risk of alarming some of my readers, who will perhaps throw my book aside as a mere maze of dry technicalities.

This great bi-annual bird-movement commences as early as February; but the initial stages of the vernal immigration on the moorlands are all but imperceptible. During the cold and wintry months of February and March, very large numbers of birds, many from considerable distances, keep quietly arriving, day by day, and distribute themselves over the moors. In the aggregate their numbers are immense, but when spread over so wide an area, their advent is inconspicuous and may easily be overlooked, especially as, on first arrival, the strangers are apt to be shy and distrustful—they have not then thrown off their wild character, or assumed the careless disposition of spring. It requires, indeed, close observation to detect the progress of the metamorphosis which then occurs. Observe those half-dozen Golden Plovers scattered over a moss-flowe high out on the fells: it is the middle of February. Well, surely, there is nothing remarkable in that? There are a couple of hundred of them in the low-lying pastures only a mile or so away. Quite true; but those hundreds in the valley are merely the normal winter stock: this handful on the hills is the vanguard of the invading army from southern lands which mean to spend the summer there.

The following list gives a rough outline of the various birds which come to breed on the Northumbrian moorlands, together with the approximate (average) dates of their arrival :—

Peewit (irregular)	February, or even end of January.
Golden Plover (irregular)	February.
Skylark	February.
Curlew	February.
Pied Wagtail	February (end).
Titlark	March (early).
Stockdove	March (middle).
Grey Wagtail	March (middle).
Wheatear	March (end).
Ring-Ouzel	March (end).
Redshank	March (end).
Blackheaded Gull	March (end).
Dunlin	April (early).
Hirundines	April.
Cuckoo	April (end).
Sandpiper	April (end).
Willow-Wrens	April (end).
Landrail	May (early).
Nightjar	May (middle).

Several of the above birds, it will be noticed, belong to species which are found in this country at all seasons of the year. As such, they may therefore be thought out of place in a list of *migrants*. But they are not so. Migration is far more general and universal among birds than is commonly, or popularly, supposed. It is, of course, a matter of common knowledge that such birds as the Swallow, the Cuckoo, and the Willow-Wren, are distinctly foreign migrants. Their summer and winter haunts are far apart, separated by a belt of sea and land: consequently their re-appearance here every April, after a total absence of six or seven months, is markedly conspicuous, and appeals at once to eye or ear in an unmistakable manner. Their annual migrations, in short, are so patent as to be obvious even to the least observant.

But there are other classes of wanderers whose movements are not so conspicuous; but which are, nevertheless, quite as strictly migratory in their habits. Thus there is the case of birds whose summer and winter limits may be said to *overlap*. Such birds are, of course, found permanently within the boundaries of the *overlapping* zone, as shown in the rough diagram annexed:—

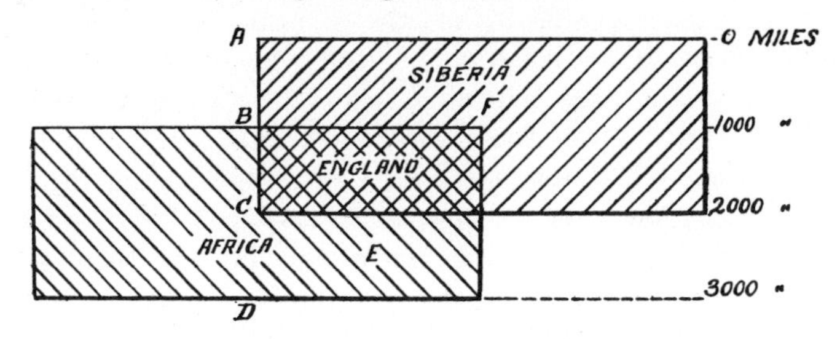

The upper oblong represents the habitat of any given species—say the Golden Plover—during *summer*; the lower oblong its habitat during *winter*. Assuming that the annual range of each individual bird is approximately equal, those breeding in summer at A would winter at B—the two most northerly points of the respective areas. The intermediate birds summering at B would likely winter at C; while those at the latter point would move on in winter to D—the two latter being the most southerly points in the two areas. In the overlapping zone (the doubly crossed portion from B to C) there will obviously be found Plovers at all seasons of the year. But they are clearly not the same individual birds. Those individuals which occupied this area in summer will be wintering—say 1,000 miles south—at E: while their vacated places are re-occupied by others which had passed the summer 1,000 miles north, at F.

Now such birds are clearly quite as much *migrants* as are the Swallow and the Cuckoo. If one happened to live in Siberia or in Africa, there would be no difficulty in recognizing the fact: but to those happening to live as we do, in the central area, where the summer and winter ranges overlap, it is less patent, and easily overlooked. It is a prevalent mistake to regard birds of this class as *resident* (or non-migratory), and this is a point I wish to elucidate.

Take the Curlew as an example. Probably nine people out of ten—shore-shooters and others accustomed to seeing them—will tell you that the Curlews are on the moors all the spring, and on the coast all the winter. This is perfectly correct; but the inference deduced therefrom, that Curlews are non-migratory birds, is wholly wrong. For if the observations be carried a very little further, it will be found that *long after* the nesting Curlews have taken up their summer quarters on the moors, there *still remains on the coast* the full winter stock of Curlews in undiminished numbers. There is, in fact, at that period—say March and April—a *double* stock of Curlews in this country.* There are, on the hills, the newly-arrived migrants from (say) the Mediterranean; and on the coast, our own winter Curlews still linger, waiting till their instinct tells them that the northern lands are clear of snow, and ready for their reception.

The case of the Golden Plover is analogous. Our winter Plovers can still be seen frequenting their ordinary haunts in large *flights*, long after the breeding *pairs* have settled down in their summer residences. Indeed, as the Golden Plovers breed rather earlier than the Curlews, it is quite a common occurrence to find the home-breeding birds (which have wintered in the south of Europe) *sitting hard*, or even hatching, at the end of April, *and at the same time and place* to observe flights of northward-bound Plovers preparing to depart, and which a few weeks later will be nesting perhaps in Siberia. These latter may be distinguished (apart from the fact of their being still in *flocks*) by the much more perfect development of the black breast of summer—a feature I intend to refer to later.

Similarly, though the extent of their range may differ, the Skylark, Titlark, Black-headed Gull, Redshank, and other birds included in the above list are to be found permanently throughout the year at one point or another of the country;

* I had the Curlews carefully watched on the coast last year (1887). They remained on the sands all through the months of March and April, the last being observed on April 30th. But on the moors they had appeared as early as March 4th, and by the end of April some had commenced to lay.

yet they are all distinctly migratory, and it is doubtful whether any of the particular individuals found in winter on our fells, fields, or shores ever remain to breed here during the following spring, or *vice versâ.*

The subject of migration, with the cognate subject of the geographical distribution of species, has of recent years received close attention from scientific ornithologists; and a flood of new light has been thrown upon the question by their researches, and especially by the systematic observations recorded at the various light-stations around our coasts and elsewhere.* But in a popular sense the subject is but very superficially understood, and this must be my excuse for endeavouring to lead my readers into a brief consideration of what I trust they may find an interesting study.

Recent investigations have clearly shown that migration is vastly more extensive than was formerly supposed. The further it is studied, the more general appears to be its scope, and the more universal the instinct in birds to migrate. Comparatively few species remain absolutely stationary throughout the year. A reference to the migration reports of the British Association shows that many of our common small birds—such quasi-domestic species as Thrushes, Blackbirds, Starlings, Larks, Rooks, and others not usually regarded as migratory—cross the seas at certain seasons in astonishing numbers, perfect streams of bird-life. The greater portion of these winged hosts is, no doubt, directed upon Continental Europe, but a due proportion reaches our shores, including members of almost every genus, and indeed of almost every species. Very few species, as already remarked, are entirely stationary; though some have very restricted ranges, while others (perhaps of closely-allied genera) are cosmopolitan in their wanderings; and of some eminently migratory species there also exist locally-resident races, which to that extent vary the general rules, and add an element of pleasing complexity to the study.

As I have pointed out, we have in our latitude a numerous

* These "reports," nevertheless, appear quite as remarkable for what they omit as for what they include.

class whose annual movements it is difficult to ascertain with precision, owing to the overlapping, in our country, of the belts or zones of land which form respectively their summer and their winter quarters. In order to arrive at a general understanding of the movements of such birds, it is necessary to ascertain their geographical distribution at the different seasons. In other words we must go outside our own country—often far outside it—in order to ascertain the limits of their summer and winter ranges.

Thus if, for example, we find that a given species is spread during summer over the belt of land from Siberia to Northumberland, while its winter range extends from Shetland to Morocco, it follows that the average annual range of the species in question amounts to some 2,000 miles. This of course assumes—and it appears to be a reasonable inference—that the range of each individual is approximately equal to that of the general body. Regarded in this light, it is easily demonstrable that many (or most) of the species which are popularly regarded as resident Britons, are in reality foreign migrants to the extent of a thousand, or perhaps two thousand miles twice every year. Many people will perhaps find it difficult to believe that their common homely Thrushes and Starlings are quite as strictly (and almost as extensively) migratory as are the Swallow and the Cuckoo; but the logic of ascertained facts shows that this is unquestionably the case.

Now, why do birds migrate? The question, at first sight, appears a simple one, and several answers will probably at once suggest themselves. In reality, however, it will be found a many-sided biological problem, and one of no small complexity and mystery.

Suitable climatic conditions, temperature, food-requirements, and distribution in proportion to food supply, are among the more obvious answers to the question. These and similar circumstances *influence* and, no doubt to a great extent, *regulate* migration. But, on examining a little more deeply into the subject, it is clear that, though in many cases they form main regulating factors of migration, yet they are not its primary cause.

Thus, with regard to food requirements, it is obvious that, when birds of a given species are found permanently inhabiting a certain area at all seasons, the natural or climatic conditions of that area do not render migration imperative. Therefore, when it is seen that large numbers of such birds *do* migrate, and traverse perhaps great distances, it is clear that any hypotheses based on considerations of temperature, climatic conditions, or the like, must be abandoned. Possibly such movements may be dictated by the quantity (rather than by the quality) of the food-supply; for, though a tract of country may be well adapted for and inhabited by a certain number of such birds, yet if an enormous additional influx of foreigners be suddenly thrown upon it, the area becomes overstocked, its resources unequal to the increased demands upon them, and a proportionate exodus or redistribution of population necessarily takes place.

All these, however, and cognate circumstances can only be regarded as regulating factors—they are not the first cause of migration; and there are, moreover, many cases in which no such factors appear to operate.

Many theories in explanation of the migratory instinct have been advanced by speculative naturalists. Some of these are, at any rate, very ingenious; but, resting on no solid basis, partake more of a poetic than of a scientific character. Indeed, this erection of highly imaginative hypotheses, in support of which it is so easy to collate a quantity of what looks like circumstantial evidence, but which are so incapable of direct proof, is of dubious utility. No doubt causes can be assigned to every effect, a reason to every fact; but it is perhaps wiser to admit that, with our finite knowledge, there yet remain things which cannot be explained.

One theory, however, has always appeared to me to stand on somewhat more tangible foundations than the rest; and, though it may be at the risk of appearing to neglect in practice what I have just preached, I will venture very briefly to refer to it.

The theory I allude to is that of the Polar origin of *all* life. This subject is now being discussed by Dr. Tristram in relation to its bearings both on the present distribution and on the migrations of birds (*Ibis*, 1887, p. 236, 1888, p. 204). With respect to the first-named point, all the salient facts relating to present generic distribution, gleaned from the four quarters of the globe, are adduced by the accomplished Canon of Durham, and the various steps of evidence by which the North Polar region is shown to have been the original centre of dispersal of ALL LIFE, should be carefully read by *all* ornithologists.

Beyond its general bearing on the correctness of the whole theory, it is unnecessary at present further to dwell on that section of the subject; but I will endeavour, in as few words as possible, to indicate the influence of the Polar theory on migration.

It must, in the first place, be granted that our globe was "in the beginning" a molten, lifeless mass; that during unknown æons it was gradually cooling preparatory to the reception of life. So much I assume. But the cooling process would clearly not proceed with equal speed. Those portions of the earth which are furthest removed from the power of the sun, and which most rapidly radiate their heat into space, would necessarily be the first to cool, and therefore the first to become capable of maintaining life. These colder portions (*provided* that the axis of the globe has not materially altered in relation to the sun) would be the Polar regions—Arctic and Antarctic. That the North Polar region has so passed through all the stages from intense heat to their existing intense cold, is evidenced by their geological record. In the interval—the wide interval between molten heat and "eternal ice"—the Arctic lands have passed, stage by stage, through every gradation of climate, and have been, at one period or another, adapted for every form of life. Spitzbergen and Franz-Josef Land have once luxuriated in the profuse wealth of plant-life of the carboniferous epoch. Incidentally I may mention having myself witnessed in the first-named ice-bound land palpable evidence of that period of "grass and herb yielding seed, and fruit tree yielding fruit"—though at the present day not a tree or a shrub exists there—and a small series of fossils brought home from

Spitzbergen were identical with those of our own coal-measures of Durham and Northumberland.

The whole theory obviously depends on the presumption that the earth's axis has remained comparatively stationary. But has this been so? This, again, is a problem the answer to which depends on a consideration of an intricate congeries of facts and forces all of which must be studied and their effects calculated. Nor can they be examined separately; they must be regarded as a great moving whole, a vast aggregation of forces acting and re-acting on each other with ever-varying results. The whole system on which the earth moves through space, the effects upon it of attraction, counter-attraction, and even such complexities as the precession of the equinoxes, all have their bearing on the question. It is, however, sufficient here merely to name such awe-inspiring topics, and to add that a consideration of them appears to justify a conclusion that the earth's axis has *not* materially altered in relation to the sun.

There is abundant evidence of tropical periods at the Pole; but no trace of glacial conditions in the tropics, nor indeed further south than the Continent of Europe. The Arctic regions have from time to time occupied somewhat different zones of land—they have extended as far southward as the Pyrenees, but not much, if at all, beyond. A Polar variation to that extent is explained by the phenomenon known as the "nutation of the earth"—that is, the oscillation of its axis accordingly as the attraction of the sun, and the counter-attractions of various other planets alternately predominate. Beyond these limits, the position of the axis appears to have been stationary—that is, it has not altered to a degree which would be destructive to the theory of the Polar origin of life.

Granting, then, the substantial accuracy of what I have feebly attempted to describe, it follows that the Polar regions (I refer to the Northern Hemisphere) would be the first spot on the globe adapted to sustain life: that they were, at first, the cradle of all life; and afterwards, as the cold gradually intensified, the centre of dispersal whence the various forms were distributed throughout the world, as its various portions in turn became adapted to their requirements.

Viewed in this light, the great migratory tendency towards the North becomes explicable and comprehensive enough. It simply arises from an innate perennial instinct, which still continues to draw vast numbers of the feathered tribes towards the point which was originally the universal home of all. It is an invariable rule that all birds do breed at the most northerly and coldest points of their annual range. In the Northern Hemisphere, the tendency to move northwards in spring is all but universal; and, as already pointed out, there are, in many cases, no visible or existing reasons, climatic or otherwise, which make such movement imperative.

Whatever may be the primary cause of migration, whether it arises from the old-time instinct I have alluded to or otherwise, it is at least certain that it is a deeply implanted and widely spread impulse throughout the feathered tribes. On referring to the foregoing list it will be seen that as early as February the influx of visitants from southern climes commences, and during that month and March the majority of the typical moor-breeding birds have distributed themselves over the hills. The Plovers and Curlew come first, followed by Larks, Wagtails, Gulls, and Redshanks, all these having northerly winter ranges, and hence comparatively short distances to come, are just what one might expect first. The Ring-Ouzel, too, from Southern Europe, follows close behind them. Of the trans-Mediterranean group, the Wheatear is the first to arrive, some weeks in advance of the main bodies of warblers, Swallows, Cuckoo, Landrail, and Nightjar. But as though to show how unsafe are any general or dogmatic rules, the Common Sandpiper, which winters in Spain, is one of the latest to arrive; and the Dunlin, which swarms on our own coast throughout the most severe winters, usually allows the month of April to begin before putting in a tardy appearance on the moors.

BIRD-LIFE ON THE MOORS IN EARLY SPRING.

FEBRUARY, MARCH, AND APRIL.

Among the few features of this wintry time of year are the movements of the Plovers and Peewits, and, in a lesser degree, of the Snipe. Of a cosmopolitan genus, one of fast and sustained flight, these birds are all the "slaves of the weather," and as uncertain. At first their erratic movements seem fairly to defy analysis or explanation ; it appears impossible to co-relate their sudden comings and goings with any visible or palpable cause. They are here one day or one week, gone the next. One winter the moorland valleys swarm with them ; another, to all appearance similar, there are none, and while here their stay cannot be relied on to an hour.

They do, however, exhibit a tolerably regular tendency to increase about the end of January, provided the season is open. These new-comers are not the home-breeding stock returning to their spring haunts, but are a contingent of the Northern race, which, during the winter, have been dispersed throughout the midlands, southern counties and lowlands generally; and which at this period, except in severe weather, are wont to foregather on the moorland hills, and remain there in large flocks till the date of their final departure for Northern Europe, at the end of April or early in May.

Those Plovers which come here to breed arrive a little later. Withdrawing from their winter resorts in Southern Europe during the latter half of February, within a few days they have distributed themselves in pairs all over the moors, going direct to the spots where they intend to nest. By the middle of March this nesting section of the Golden

Plovers are all localized in pairs on the high moors, though they will not have eggs till a month or five weeks later. These birds are already at that date almost as black on the breast as our local Plovers ever become, for they never acquire the full black breasts invariably depicted in books, but which are only attained by the more northern races. The Northumbrian Plovers at best are only marbled. Even in Shetland, and in Southern Norway, I have noticed a much closer approach, in this respect, to the full typical development, and the examples obtained by my brother Alfred, in Finmark, were perfectly black beneath. This more complete assumption of the typical summer plumage in proportion as one goes further north is a remarkable, and appears (where applicable) a tolerably constant feature in ornithology. The inference as to the Polar origin of such species is irresistible. Thus the Bramblings (*Fringilla montifringilla*) shot by my brother in 70° north latitude, were markedly more perfect in the glossy blue-black of their heads and shoulders than others obtained on the Dovre-fjeld (lat. 63° N.) and on the Sogne-fjord (lat. 61° N.) at a corresponding season.

Again, what other birds of the known world attain so complete and perfect a summer-transformation as the hyperborean trio—the Godwit, Knot, and Curlew-Sandpiper ; the only species (with one exception) whose breeding-places are yet undiscovered—hidden amidst the inaccessible penetralia of the Arctic zone ?

In February the Curlews return ; and very welcome is the first sound of the wild long-drawn rippling note, and the first sight of the shapely clean-cut form sailing across the dark heather. Their arrival has occurred as early as February 5th, and as late as March 11th, the average being about mid-February. In very stormy seasons, when the fells are buried in snow, the Curlews delay their return till the snow has melted ; as in 1886, when none appeared on the moors till March 19th. These Curlews also are travellers from afar ; they have come from Spanish marismas, from African lagoons, and from the shores of the Mediterranean. The Curlews of our own coast do not breed here ; they remain on the oozes and sand-flats all through the months of March and April, and retire to breed in more northern lands in May.

In the latter half of February occurs the vernal influx of Skylarks, varying in date according to the character of the season. In mild winters a few of these birds remain in the neighbourhood of the moors all the winter. That of 1885–6 was a rather remarkable instance, and illustrates how completely ornithic instinct is sometimes at fault in forecasting the weather. Throughout the mild months of December and January, Skylarks had been unusually numerous; and their numbers received a great accession in February. On the 7th of that month some had even commenced to sing; but the sudden and memorable snowstorms of March 1st and following days, which buried North Britain under many feet of drifts, isolated towns and villages, and swallowed up whole trains on the railways, most effectually broke the spring-dreams of the little songsters. They wholly disappeared until the snow melted on March 19th and following days.

February 22nd is the date on which, in three consecutive years, the Pied Wagtail has made its appearance, and in a fourth year it was only one day later. The Wagtails are hardy birds; for insect-feeders they arrive singularly early. Indeed, in mild seasons, it is not unusual to see some of the Grey species (*Motacilla boarula*) daintily running about the burn-sides and shallow water in mid-winter. There occurs, however, a visible accession to their numbers, by migration, during March.

The above comprise all the species whose advent I have been able to observe up to the end of the month of February.

Early in March the Mallards and Teal return to the moorland loughs. Where they have been since December does not appear; but, even in open winters, it sometimes happens that we have hardly any ducks at all on the higher moors, except, perhaps, a few Golden-eyes at intervals.

March 8th.—The Ravens at ——* have five eggs,

* With young Ravens at half-a guinea apiece, and the insatiable—aye, insane—greed of "collectors" for *British-killed* specimens, it is

though the hills are enveloped in snow. There are now only a few spots remaining along the Borders where these fine birds are allowed to nest. Several of their former ancestral strongholds now only retain such names as Ravenscleugh or Ravenscrag to connect them with the memory of their ancient tenants. The Raven is gone, and his place is often usurped by a swarming colony of impudent Jackdaws—in the aggregate, a hundred-fold more mischievous than the solitary raven.

The Titlark is another of the spring migrants to the moors. Early in March I have noticed hosts of them on passage across the lowlands at Silksworth (county of Durham), but it is nearly a fortnight later before they appear, on an average, on the Northumbrian moors. These birds, as well as the Skylarks, Song-Thrush, Starling, and other common British species, are extremely abundant in winter in Spain, but do not remain there to breed, leaving Andaluçia at dates corresponding with their reappearance here.

Among the earliest birds to commence nesting are the Owls; in some large woods which I rented, the Long-eared Owls (*Strix otus*) were rather numerous, and exhibited some peculiarities of habit which are worth recording. Never troubling to undertake the construction of a nest for themselves, they rely on forestalling some more industrious architect: one pair (on March 19) were commencing to sit on five eggs, laid in a nest which was built and occupied the previous year by a pair of Sparrow-hawks. The Owls have played their rivals this trick several times. The Hawks are a month or more later in commencing domestic duties, and when they return to their abode at the end of April, they find that they have been forestalled by several weeks, and that their patiently-constructed platform of spruce-twigs is then fully occupied by a family of large-eyed, hissing, and staring young *striges*. The Hawks

wiser to omit names. If "naturalists" *must* all have collections, why cannot they be satisfied with the beautiful specimens which are so easily procurable from northern or eastern Europe, instead of hastening the extirpation of this, and other scarce indigenous birds, by placing a high premium on their heads?

do not appear to resent this usurpation—must, indeed, admire it, for they one year themselves proceeded to usurp the nest of a Wood-Pigeon close by, and laid their eggs alongside the Cushats' pair. This year they built another nest in a spruce-fir, not a hundred yards away, and on May 4th had seven eggs. The Owls at that date had two large young, one already able to fly, besides three addled eggs, in the Hawks' former abode.

A peculiarity in the habits of these Owls (*Strix otus*), after the breeding season, deserves a remark. As soon as the young were fledged the whole of the Owls associated together, perhaps three or four broods, old and young, into a single family, and chose a thick, black Scotch fir for their abode. Here they all passed the day. To this particular tree the whole of the owl-life of those woods resorted regularly at dawn, and in it slept away the hours of daylight, hidden amidst its deep, evergreen recesses. At the particular tree of their choice—it varied in different years—the Owls could invariably be interviewed, during the summer and autumn, though, to a casual eye, it was difficult amidst the deep shadows of the foliage to distinguish the slim brown forms pressed closely against the brown branches of the pine. Towards dusk their awakening was notified by the querulous, cat-like cry; ten minutes later, their silent forms appeared outside the wood, and after a few rounds of preliminary gyrations, it was dark enough to commence operations in earnest. During the nesting season the old Owls have another cry—not unlike the petulant barking of a spoiled lap-dog.

In the larger woods and those of older growth we have the big Tawny Owl (*Syrnium Aluco*). This is the species whose startling cries often make night hideous—or interesting, according as their melancholy cadences affect the fancy of the listener. The Tawny Owl nests almost as early as the Eared species, having eggs by March 25th; it does not, however, like the latter, require any nest, properly so called, but lays its eggs in some hollow tree. If, in the wood they frequent, there exists an old tree, rotten and hollow to the stump, it will in all probability be selected year after year for their nest, the eggs lying on the bare scraps of "touch-wood." The hole or slit used for ingress and exit may be three or four feet above the eggs, and, perhaps, not over four inches in width; and how so large a bird gets comfortably up and down the narrow vertical tunnel is not easy to see; but this is a problem with most hole-breeding birds.

Although it is a thousand pities ever to kill one of these harmless birds, yet if any one has an opportunity of handling one newly-killed he may utilize the poor victim by examining the extraordinary formation of an Owl's ear. I will undertake to say it will both surprise and interest. By

THE DIPPER.

separating the long feathers immediately behind the facial orbit, what can only be described as a huge gap in the bird's head will be observed. So entirely disproportionate does this chasm appear, that one is inclined to wonder if the bird has met with some accident. The skin of the head is loose, and unattached to the skull for some distance round the aurital cavity; consequently the whole terior portion of the cranium, together with the reverse, or inner side of the bird's eye, is fully exposed to view.

Not many birds care to undertake the risks of domestic

affairs during the stormy month of March. Perhaps the only ones besides the Raven and the Owls are the Heron, the Rooks, and the Little Dipper. A hardy sort is the last-named, and no one who has heard them in full song during the severest weather of mid-winter, or watched them blithely diving under the fringe of fast ice along the burn-sides, when the temperature is but little above zero, would be much surprised if they commenced nesting at Christmas. Their most favourite site is in the *linns*, or small waterfalls, where a hill-burn comes tumbling over an exposed ridge of rock. Many of these linns, with their shaggy fringe of gnarled and

HOME OF THE DIPPER..

lichen-clad birch, heather, and bog-myrtle, are among the wildest and most lovely nooks of the wild moorland. There, in an interstice of the moss-grown rock, half overhung by ferns, and all but indistinguishable from its environment, is cunningly inserted the great round nest of green moss, in the very spray of the falling water. The outside of their home is splashed and wet. The old birds have to pass, to and fro, through the fringe of the cascade, but that is just what these little amphibians like, and hardly a linn but has its pair of dusky, white-throated tenants. Other nests are fixed on some big boulder islanded in mid-stream, and one on a thick, gnarled branch overhanging the water.

The interior of the nest is dry and warm enough, and the eggs number five or six, pure white, but showing a pretty pink blush when in a natural state. The young are on the wing in April (late broods up to June) and from the first take lovingly to the water—diving like water-rats even before they can fly.

The last week of March witnesses the arrival of several of the typical moor-breeding birds. The Stock-Dove and the Grey Wagtail have returned by the middle of the month; and its penultimate days bring the Wheatear, the Ring-Ouzel, the Redshank, and the Black-headed Gulls; possibly, also. the Dunlin, though I have not chanced to see these till April. The first three are very regular, as the following records will show:—

	Earliest date.	Latest date.
Wheatear arrives	March 23rd.	March 31st.
Ring-Ouzel „	March 24th.	April 2nd.
Redshank „	March 11th.	March 27th.

The Redshanks come by night, and the first intimation of their advent is often the well-known triple note, heard far up in the dark skies. During the second half of March, the Black-headed Gulls (*Larus ridibundus*), among the most conspicuous ornaments of the hill-country during the spring and summer months, straggle up somewhat irregularly. At first they keep the lines of the river valleys and main "water-gates," but by the end of March have reached their *incunabula* among the moorland lakes. The Stock-Doves, too, come irregularly—their earliest date being March 7th, and the latest April 1st. To the moors, these small wild pigeons are *exclusively* summer migrants. None are met with during the autumn or winter—the latest I have observed was on August 18th—although of recent years Stock-Doves have become tolerably numerous at that season throughout the *lowlands* of the northern counties (See chap. on "Wild Pigeons"). On the moors, they appear annually during March, and at once locate themselves at the crag where they intend to nest.

March 20th.—A Missel-Thrush commenced laying, though there were 7° frost at night. Rooks, Herons, and an

OLD BLACKCOCK AND GREY. (APRIL.)

To face page 24.

occasional Wood-Pigeon begin to lay about the 25th. March is usually a good month for Jacksnipes. I have several times met with little parties of five or six together, both on the moors and the lowlands. These are evidently resting on their passage northwards, for, even if not shot, none are to be found the following day.

March 31*st.*—Killed to-day the first adder; another on April 6th. They are common on the moors throughout the summer months, living on mice and small birds, and often met with while grouse-shooting. On September 25th (1881), I killed an adder of a peculiar warm reddish hue—quite different to the ordinary colour. It was gliding down a steep slope, moving on the top of the heather, and contained three whole field-mice. This adder was also the latest I recollect seeing; they go into winter-quarters about the end of September.

The Ring-Dotterel (*Ægialitis hiaticula*), in this country, is usually marine in its haunts, and we have never known of its breeding inland; yet at the end of March they are not uncommon upon the wide haughs of Upper Coquetdale, (25 miles from the sea), associating with Redshanks, Peewits, and an occasional Dunlin.

April.

The most thoroughly characteristic sound of the moorland valleys at this season is the peculiar love-song of the Blackcock. One hears everywhere the strange low bubbling note, and presently detects its author in the form of a black and white spot, far out in the centre of some wide pasture, or on the rush-clad slope of the hill. Around the excited mass of black and white feathers sit his consorts—half-a-dozen Grey-hens, some picking at the rush-seeds, others preening themselves, or enjoying a quiet siesta; but all supremely careless, and to all appearance unconscious of the elaborate demonstrations which are being performed for their behoof. Altogether, it is an extraordinary spectacle, and one that is somewhat involved in mystery. During the months of March and April the performance is incessant;

even in February it is frequently audible on fine mornings, especially about daybreak. We have noticed it as early as January 30th, but at that early period it is confined to the early hours of the day, and sometimes attended by terrible combats between rival monarchs. By the middle of April, however, the extreme virulence of their animosity towards each other appears to have cooled down, and small parties of Blackcocks may be seen amicably feeding together. On April 14th, in a small enclosure of meadow-land, on Redewater, no less than twenty-one were thus assembled. There were no Grey with them, yet several of the biggest old cocks were walking about (feeding) with all their spreading tail-feathers erect, and partially distended, as though that position was chronic at this season.

REDSHANK IN APRIL.

The Peewits are now laying in earnest; we found an exceptionally early nest on February 28th, but the first week in April is the best time for gathering their eggs. A friend recently asked me how it was that Plover's eggs can be bought in London all through March; if that is so, the question must be put to some one other than a naturalist. By April 10th, the Redshanks are all localized and settled down in pairs about the low rushy meadows, and stagnant backwaters along the riverside, where they mean to nest. Their actions are now very graceful as they wheel round overhead, alternating a rapid flight with short jerky periods, and poising in mid-air on wings curiously held bent beneath them, and pointing stiffly downwards. The young Herons are now nearly half grown; the Rooks "chipping," and Corbies (*C. corone*) beginning to sit. The latter breed in nearly every scattered clump of pines, often far out on the moors. Reed-Buntings have appeared in full summer plumage, and the pine-woods swarm with Goldcrests.

April 13*th*.—Started this morning early to fish some streams on the extreme Border—a long drive. The weather bitterly cold, snow falling as we started, and at intervals all day. There is but little sign yet of the return of spring, either in the weather or the vegetation, in this wild country. The trees remain bare and naked, and the moors barren, bleak, and wintry—none of those bright hues which

REDSHANK IN APRIL.

in autumn adorn the flowes and mosses; nor any leaf as yet on birch or alder, though the birch-woods at all times show a warm purplish gloss on their dark masses, like the bloom of a ripe grape. This effect is produced by the deep claret-hue of their twigs, and not by the rising sap.

The snow lay in great ominous patches along the heights of the Carter and adjoining fells, and several times during the day furious snowstorms swept down the dale, driving before a north-easterly wind; yet the trout rose freely and well to the fly, and we creeled 4½ dozen nice fish (2 rods), though it was frequently impossible to keep sight of the flies amongst the thickly-falling flakes. Next morning the ground lay white as winter; not a trout would rise, but I got a clean-run bull-trout of 2 lbs., the first of the season. Many Lesser Black-backed Gulls frequent the Border rivers at this period.

On April 14th, my brother A. found a nest of the Grey Wagtail in a crag on Coquetside; a month later I found another, with three new-laid eggs, on the Dunshiel Burn, and others at intermediate dates. The Wagtails thus begin to lay before the bulk of the other insectivorous birds have arrived. The first nest of the Pied Wagtail was found in an old stone-dyke, on the 17th; another, in the burn-side, on the 21st. The Yellow Wagtail does not frequent the hill country.

Year after year, at this season (mid-April) I notice, while fishing along the burns, considerable packs of Golden Plovers still frequenting the haughs and lower slopes of the hills, though some of the local-breeding Plovers will already have commenced laying higher out on the heather. These packed birds are visibly blacker underneath than the nesting pairs; and one frequently hears far overhead their loud and wild spring note—"Tírr—pēē—yŏu"—a note which the former section have now almost ceased to utter. It is only heard when the birds are on wing, and high in the air. The breeding Plover's note is now confined to the single plaintive whistle, and their peculiar rippling song or warble. This latter, which is quite indescribable, is a sort of joyous note of courtship, corresponding with the "drumming" of Snipe and Peewit—indeed, many of the wild birds have an analogous note at this season. Among the bogs and mosses, the Snipes course high overhead, often a dozen or more at a time, and the strange bleating cry comes down from mid-air, alternating with the sharp metallic "chip, chip," when flying free. Dunlin, Redshank, and Curlew—indeed, most birds of that class—have dual notes; these are their ordinary means of alarm or communication; and besides, the note indicative of the exuberant spirit of the vernal season.

Snipes only drum head to wind, and when falling. On the opposite page is a diagram of the flight of a drumming Snipe.

The drumming is produced only during the *dotted* periods of the bird's flight, and is clearly attributable to the wings (not the voice), for the key changes with the alteration in the bird's course in the air, as at (A). Snipes commence to drum about the middle of March. Another of the seasonal signs, which, one after another, spring into being, is the busy hum of the humble bee—and few sounds are more welcome, or more pleasing to the ear than this, when first heard on a sunny morning, about mid-April. There is a thoroughly summer-like ring about it, which serves as much as anything to mark the change of the season, and the termination of wintry conditions.

By the middle of April, many of the typical moor-birds commence their domestic duties in earnest; but the 20th of the month I regard as, in average seasons, the standard

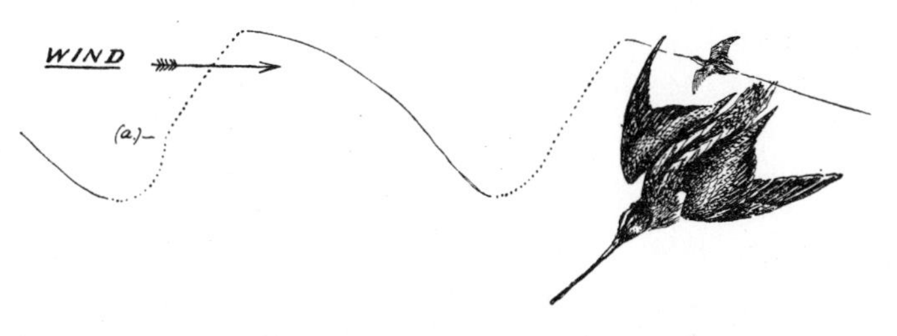

date of laying for Grouse, Mallard, Golden Plover, Snipe, Pied and Grey Wagtails, and Stock-Dove. The Curlews, Gulls, and Black game are a week or ten days later. I have seen nests of the first-named group as early as the beginning of April; but this is exceptional, and, as above stated, the 20th of April may be regarded as the average date for completed clutches.

The Curlews are not at all particular as to site; their nest is usually high up on the hills, but grass or heather, long or short, bare or dry ground, or bog—all seem to suit them alike. Even when the nest is among long heather, there is no premeditated attempt at concealment. The old Curlew relies on her watchful nature and keen eye for safety, and rarely sits close when danger threatens, however distant; though, from her size and light colour, it is not difficult to find the nests, when one knows how to look for them. The four eggs are laid in the final days of April, one or two being often unfertile. The young Curlews do not leave the nest immediately on being hatched, as the young of most of this class of birds do; for, though they may never be found actually in the nest, yet they will be lying hidden in the heather close by, having just slipped out on the approach of danger. This, of course, only applies to the early days of their lives.

Golden Plovers seldom or never nest among covert—*i.e.*, the nest is on the shortest grass or heather, often on perfectly bare ground. There is no attempt at concealment. On being approached, one Plover will rise straight from her eggs, a couple of hundred yards away; another slinks off, creeping away unseen through the heather; at other times, though more rarely, she will rise off her eggs at one's feet, even when fresh laid. The young run as soon as hatched, but are long in acquiring the power of flight, and retain the golden down on their necks when full-grown, as any grouse-shooter can see in August.

Peewits breed in thousands on the lower grounds, and not on the high moors preferred by Curlew and Plover; Snipes at all elevations, on hill or valley. Their nest is always well concealed under a tuft of grass or heather; and the old bird sits close. Snipe are very irregular in their dates of laying; I have found young ones unable to fly on August 12th, and, on the other hand, have known of a nest as early as 19th March, and of young Snipes on the wing in the last week of April.

The Stock-Doves have laid their two eggs by April 20th, but, like all the Pigeon-tribe, they compensate for their small broods by nesting continuously all through the summer, often having fresh eggs at the end of June. They breed in the crags, a very few birch twigs or bits of heather serving for a nest. In vertical crevices, however, a good deal more material is required to obtain a foundation, unless the Jackdaws have previously filled the hole with sticks, as their habit is. On May 7th, we found a nest in a different position—under an immense boulder at the Cloven Crag, a mass of tumbled rocks flanking a singular cleft in the hills. The eggs were lying on the bare peat.

The 20th April is also the date when the Sandpiper is due to re-appear on the burns and river-sides.

Though the weather is still bitterly cold, and the sting of the east wind has lost none of its marrow-piercing venom, yet the packs of Golden Plovers have now entirely disappeared from the low-lying grounds. None are to be seen here to-day (April 27th), where a fortnight ago there were hundreds. Instinctive perception never fails as to dates; they know their appointed season has arrived, and they have gone on northwards to the wastes of Northern Europe and Asia. The Curlews, too, have disappeared from the coast. My winter-puntsman wrote me that April 30th was the last day on which he observed them in any quantities on the sands, and curiously, on the same day we found the first nest, with four eggs, on the moors. The above dates correspond with the appearance of these birds in the far North. I have a note of the arrival of Plover in Nordland (at Langnœs, Tromsö), on May 12th, and a few days later a Curlew's nest was found, which was considered there exceptionally early.

OLD BLACKCOCK "IN PLAY."

SOME NOTES ON THE GAME-FISH.

During the month of April, in average dry seasons, the Redewater (and other rivers) are full of salmon and bull-trout kelts, unable to get down to the sea for want of water. Many of these fish are in the last stages of disease, long, lank, hungry-looking brutes, disfigured with great white leprose spots, especially about the head, and lying inert and listless in backwaters and burn-mouths. For these a flood-water would avail nothing. They are too far gone, and the sand-banks are strewn with the bodies of those already defunct—choice morsels for the Corbies. Sometimes one of these dying kelts gets into a stream too rapid for his strength, and is carried down "all ends up"—tail first, head first, or sideways—and one sees his white belly glint as he tumbles over and over in the current. Or in wading in the dead water, one, perhaps, almost steps on to one of these poor moribund fish—it is rather surprising, on looking down for a fresh foothold, to see a great salmon's tail wearily fanning the water within a few inches, and the fish apparently quite unconscious of one's approach.

Up to the middle of April there is, probably, not a single clean-run salmon in the upper waters. But they swarm with smolts, both the game-looking silver-clad fish, which I suppose is the young of *salmo salar*, and the coarser-looking, "finger-marked," trout-like kind, with large red spots on his flanks, which is probably the young bull-trout. The silvery fish are all gone by the end of April; but the others remain in the river all the spring, and by June some have grown to a fair size. There are also thousands, or tens of thousands, of the small "par," or "rack," whose species is probably undefinable precisely, by any one. By the law of the land, it is illegal to take any of these various fish, and penalties can be enforced against any one killing them. The utility of this appears to be, at least, questionable. A spawning salmon or bull-trout produces some 800 ova to each pound of her weight—(say 16,000 ova to a 20 lb. fish), and of this enormous number, a very small—quite inappreciable—percentage is ever intended by Nature to reach maturity. From the first they are designedly the food of many other creatures. Trout, eels, the larvæ of water insects, and even their own kind devour them in thousands from their very birth. Then on their development into fry, they are preyed on by other fish, Herons, and diving birds all the way down to the sea. Their arrival in salt water is awaited by a host of fresh enemies—cod, coalsay, Cormorants, Goosanders, and divers of every kind, and no one knows what else. During May and June, the rock-codling, poodlers, coalsay, and lythe, appear to live almost exclusively on smolts, from half-a-dozen to a dozen of which are often found in a single fish, and the total numbers of these marauders is legion.

Probably, not more than perhaps four or five smolts in every 10,000 was ever intended to survive all the manifold risks and enemies it has to encounter, and the extra risk by rod and line would probably be an inappreciable factor. The foregoing remarks are suggested by seeing a blue-coated, silver-buttoned public functionary zealously pursuing a few bare-legged, terrified little urchins who have dared to attempt the capture of a two-ounce par.

The period when salmon do want protection is in late autumn and winter, when they run up the small burns, and into the shallow waters of the stream-heads to spawn. Like all other wild creatures at their corresponding season, salmon at that period become wholly defenceless, and, as it were, throw themselves on human mercy and consideration. And surely, in return for the sport they give, and for their delicious flesh, they are entitled to have that consideration at the time when they cannot take care of themselves. They are surely entitled to be protected against the greedy rapacity of the poacher, and the nocturnal fraternity of the lantern and the cleek? The period of danger is comparatively short, and a dozen men employed then for a month, would be more effective than one man employed the whole twelve. Let the salmon spawn in peace; give them a fair and open run up and down the rivers, and there would be no fear for their numbers, and no necessity to trouble about a few thousand par or smolts in the spring.

One other circumstance in connection with the spawning of salmon it may be worth while mentioning here, for its local interest. The Redewater divides near its source on the Border into two branches. The main stream flows west through moor and moss; the eastern branch traverses a bare, cold rock formation of similar character to that of Upper Coquetdale, where it is so remarkable that no true salmon can be induced to spawn. This rock formation is, in fact, a spur of the plutonic rocks of the Cheviots, which terminates in the Lumsden Law, close to the head of Redewater. The water-bailiff tells me that *all* the salmonidæ take the *main water* of the Reed for spawning; *none* ever ascend the eastern branch. Thus it appears that it is this geological formation which is obnoxious to the spawning fish; it explains the absence of true salmon in Coquet, and the failure of the costly experiments and continuous efforts which have been made to convert it into a salmon river. Had the geological conditions been studied in relation to their bearing on the requirements and the economy of the fish, all this labour and expense would never have been incurred.

For the sake of accuracy, it should be added that though in Upper Coquetdale, the bull-trout spawn abundantly on the cold rock formation alluded to, yet in Redewater, where they have a choice of both stratified and plutonic rock, they invariably avoid the latter.

CURLEWS IN MAY.

To face page 34.

BIRD-LIFE ON THE MOORS IN MAY.

THROUGHOUT the Borders in spring are scattered colonies of the Black-headed Gull (*Larus ridibundus*), and the larger breeding-places are among the most interesting and animated of moorland scenes. Nearly every large sheet of water has an extensive colony: while many a remote moss-pool and nameless hill-lough has its two or three pairs of this graceful species.

There is now (May 1st) an abundance of Gulls' eggs at our lough—about 150 nests. Ten years ago there were not a dozen pairs, but preservation and the destruction of their enemies have largely increased their numbers. The nests, made of heather and dead rushes, are crowded thickly together along the lough-side, among the short heather and on the soft spongy green moss, or sphagnum. Hardly two eggs are alike, even in the same nest—a sort of *concordia discors*. If "birds in their little nests agree" (which is doubtful) certainly their eggs do not. Those of the Gulls are of every shade of greens, blues, and browns: some dark and heavily blotched, others pale and almost spotless; and the variation even extends to their shape. The full complement of eggs is three; but some nests hard-sat, or even "chipping" (on May 18th) had but two, or even only a single egg. There were no young fairly hatched out at that date, but on June 2nd, they were out in numbers, some old enough to creep away among the heather. The young Gulls at first are very pretty objects, warm yellowish-brown in colour, spotted with black, and with large eyes, full and dark: their beaks and legs pinkish, the former tipped, and the latter shaded, dusky. When fully fledged they are very prettily mottled above with warm browns and black, which contrast pleasingly with the pale French-grey and snowy-whiteness of the rest of their plumage.

The mischievousness of the Jackdaws at the hatching period is almost incredible. They constantly hover round the Gullery on the look-out for plunder; and everywhere lie young Gulls, dead, with a sharp beak-thrust through their soft stomachs. The Jacks continually engage in skirmishes with the old Gulls, and are evidently masters of the situation. We set a dozen traps near the Gullery, baited with eggs, and caught about thirty of the impudent marauders; but it was noticeable that not a single Gull came near the traps. This shows clearly that the Black-headed Gulls are not guilty of egg-stealing propensities; and keepers have therefore no reason for molesting these beautiful moor-birds. They feed on worms, slugs, &c., and come down regularly on to the grass-land in search of such food. Very different is the habit of the Black-backed Gull (*Larus fuscus*): these large and powerful birds are inveterate egg-stealers, and deserve no mercy on the inland moors. On one occasion I found one of the big Gulls floating, dead, in one of the loughs at Elsdon, choked with a wild-duck's egg stuck fast in his gullet. (See Yarrell's British Birds, 4th Ed., Vol. III., p. 627.)

It is this species (the Lesser Black-back) which is so numerous on the coast at the Farne Islands, some 30 or 40 miles away, where many thousands breed. I do not know of any nesting on the inland moors at the present day, though they used to do so many years ago. Mr. C. M. Adamson tells me, in the neighbourhood of Haltwhistle, and doubtless elsewhere also. But they were persecuted (I must say deservedly) by the gamekeepers, and are now apparently driven exclusively to the salt water. They are frequently to be seen, nevertheless, on the moors and inland rivers, especially during spring, and it is possible that some may still nest in the almost boundless tracts of moor and waste along the Borders.

The Grey-hens are also beginning to lay. One nest to-day (May 1st) in an old "scroggy" fence, had two eggs; another on the 4th, out on the open moor, had four. Though they frequent all the higher fell-ranges of the Border, yet Black-game are not so exclusively alpine in their tastes as the Grouse; the Grey-hens in spring seek the lower grounds for their nests. The situations selected are the moist and rushy slopes of the hills, where rough grass, or ferny and straggling birch-woods begin to take the place of heather; and one of their greatest strongholds is the broad zone of rough undulating prairie land, which lies subjacent to the moors proper —in short, betwixt the heather and the corn.* Grey-hens are often singularly foolish or careless in their choice of a site; one nest is in a clump of rushes immediately adjoining a stile where men and dogs pass daily; another in the bank thrown up to form a sheep-washing pool on a burn; both of which were inevitably destroyed. The Grey commence sitting about the middle of May.

In 1877, we had a severe snowstorm early in May, lasting some days. On the morning of the 4th, the snow lying 3 or 4 inches deep, I shot some Fieldfares, out of a large flock. They remained about all that week, though in a fortnight's time they would be due to have laid their eggs in Norway.

May 7th.—A single Golden-eye Duck still on the lough—another northern breeding bird. The weather is now quite warm and summer-like, and it *seems* late for so essentially a winter duck to be lingering here; but the fjeld lakes of Norway, where it probably intends to breed, are still ice-bound, and will remain so for another fortnight. This I happen to know from personal experience; my little diving friend knows it equally well from intuition or instinct. Hence it is in no hurry to be off.

We found to-day (May 7th) a Ring-Ouzel's nest on Leechope; another on the 13th, in Lanshot Scroggs, each with 4 eggs. The Ring-Ouzel is a typical moor-breeding bird; it passes through England in March, and arrives on the northern fells in the final days of that month. It nests pretty well all over the moorlands; but the majority of pairs

* To define the relative distribution of Grouse and Black-game in spring, the former may be said to nest at the highest, the latter at the lowest, points of their respective areas. Grouse in spring seek the higher ground for nesting, and in autumn, so far as they move at all, tend to shift downwards; while Black-game breed chiefly in the lowest ground of moorland character, and as the young acquire strength in autumn, tend to climb outwards to the higher fells.

select a heathery slope high out on the hills, where the nest is placed loosely among the long heather on the open moor. Others prefer more sheltered situations in an overhanging bank, among scrub or fern, or a rocky scaur. Sometimes, on the burns, the nest is on a ledge of a crag in close companionship with one of a Dipper. Outwardly it resembles a Blackbird's, but has a framework of heather stalks. The Ring-Ouzels lay 4 and sometimes 5 eggs about the first week in May; hatch towards the end of the month; and the young are on the wing by the middle of June. They remain on the moors till the autumnal crop of berries—especially the mountain ash—is exhausted, when they return southwards.

May 8th.—Cuckoos observed paired; watched them for some time. Only one calls, probably the male. A Dotterel (*Eudromias morinellus*) was shot on the moors above Wolsingham, co. Durham, by Mr Crawhall's keeper (Pattison Wearmouth) on the 14th May. Dotterels pass northward through England at this season, but are never common, and none breed in Northumberland or Durham.

The Dunlins must have eggs now—a week ago their actions showed they had already laid—but on the immense extent of ground, it is all but impossible to discover their nests. Their most favoured haunts are some wide tussocky flowe, far out on the hills, and perhaps a mile in circuit. This great flat area is occupied by perhaps but a single pair of Dunlins; hence the difficulty of detecting the exact site of the nest is obvious. To attempt to watch the birds on to it is vexation of spirit. They are so ridiculously tame, running unconcernedly around, almost within arm's length, "purring" the while in their peculiar fashion, that one imagines the nest must be close at hand. Then after lying patiently watching them, for perhaps half an hour, up goes the Dunlin with a little wild pipe, and flies right out of sight. I have seen them year after year in spots where they certainly do *not* breed, perform all their presumptively breeding antics, as though gratuitously to deceive one. It will thus be seen that though in the aggregate, a good many Dunlins nest on the Border moors, yet being scattered widely about in single pairs, they are easily overlooked.

The Redshank is another bird whose nest is rarely found on the moors, by reason of the sparse and scattered distribution of the breeding pairs, and the elaborate concealment of the nest. These do not, like the Dunlin, breed high up on the fells, but prefer the rushy fields of the lower grounds and small patches of bog.

Both the species just mentioned breed in some numbers on the great marshes of the Solway, and may there be much more readily studied than on the highlands of Northumberland. These marshes are of great extent—for many miles a dead flat, grassy expanse, hardly raised above sea-level, and intersected by muddy channels, and creeks of salt water—a very different region to that frequented by the Dunlins on the moors. I well remember one day my brother Alfred and I spent on these marshes in mid-May. It was intensely hot, and we were panting with thirst (all the water being brackish), when right before us, a mile or two distant across the flat marsh, we observed a large house embedded among trees, and having a broad sheet of water lying between us and it. It was an exquisite mirage. So perfectly distinct was every detail, that we felt inclined to hurry forward for a drink of fresh water from the lake. Whence came that image it is impossible to conceive. There was not, we knew, an object on the marsh bigger than a mole-hill, there was not a drop of water, nor was there in that direction a house nearer than the opposite shore of St. George's Channel. I have seen many remarkable mirage effects in the great flat *marismas* of Southern Spain—and once in the Arctic Seas we observed a singular appearance of rocky surf-beaten islands where none existed. Two ships which happened to be in sight, were also reproduced, inverted immediately above the actual vessels. In all these other cases, however, there was always more or less of distortion or extravagance. Here every feature was natural and defined. Never have I seen so perfect an optical delusion as that mirage on the Solway.

Though the birds were abundant enough on the Solway marshes, there are few nests so difficult to find as that of a Redshank. She hollows out some thick tuft of coarse grass, the tops of which, twined together, completely hide the nest from view. There is merely a sort of tunnel leading transversely through the tuft, which serves for entrance and exit, and her long neck enables the sitting bird to observe afar the approach of danger, on which she at once slips silently away. Mere casual search is therefore utterly useless; it is necessary that the eye should instantly detect the bird as she springs from her nest—no easy matter at perhaps 100 or 200 yards' distance, and when the air is filled with Peewits and other birds wheeling about. Then, when one does succeed in detecting the movement at the exact moment, there still remains the difficulty of marking the precise spot on so bare and featureless a place.

On the Solway the Dunlins breed in similar situations—barely above tide level—but their habits are different. They build a slight nest, like a Skylark's, but there is little attempt at concealment. They usually run from their eggs on being disturbed, and as they have perhaps gone several yards before being perceived, one is apt to be deceived in not finding any nest at the spot where the old birds, by their actions, lead one to expect it.

The geographical distribution of several species belonging to this class, during summer, is remarkably wide. In the South, both the Redshank and the Peewit remain abundantly to nest in the blazing heats of Southern Spain. In my visits to that country (Andaluçia), I have found the nests of both in plenty, and at about the same dates as at home. I even found on one occasion, a nest of the Dunlin with four eggs on a dreary flat marsh on the lower Guadalquivir—a place somewhat resembling in general aspect the marshes of the Solway. Then, in the far North of Europe, my brother Alfred found the Redshank breeding in Finmark (lat. 70° N.), and both it and the Dunlin are included by Wheelwright (Ornithology of Lapland, p. 5), among those birds which breed within the "region of perpetual snow." Their summer range is therefore pretty extensive, and their tastes in the selection of a summer residence about as diverse as can well be.

GOLDEN PLOVER—SUMMER PLUMAGE (MAY). (NORTHERN TYPE.)

To face page 40.

May 15th.—The Grouse are now hatching in all directions. This is about their regular average date. The Grey-hens are all beginning to sit.

This is also the date when the Sandpipers lay. As already mentioned, these charming little waders appear on every turn and river, about the end of April, and no bird, not even the Swallow, is more intimately associated with the return of summer, than is this cheery and graceful little angler's companion. Many of the pleasantest days at this season

ANGLERS' COMPANIONS. THE SANDPIPER.

are spent rambling along the burn-sides with one's rod, and all day long one is accompanied by the Dipper, the Wagtails, and the Sandpiper—the latter flitting along the stream just ahead, or perching on a rail or furze-bush, for ever flicking up his tail, and trilling out his cheery song.

By mid-May, the Sandpipers have laid their four pretty eggs under the shelter of some tuft on a shelving bank, among a bed of osiers, or such-like situation. Their nests being right in the fisherman's track, are often discovered by the old bird fluttering out across the shingle, with well-feigned lameness. The young, pretty little grey things, spotted with black, are hatched early in June, and most of them have left the moors before the end of July. Yet as late as June 10th, we have found nests of newly-laid eggs; these late sittings, as also noticed in the case of the Golden Plover, being often remarkably handsome and strongly-marked specimens.

May 19th.—A glorious summer day: the influence of the season is now fully developed, and the full chorus of the summer birds can now be enjoyed to perfection. Besides those already mentioned, there is now added to the orchestra the notes of all the little Warblers, the Chats, and the ubiquitous Cuckoo, and to an appreciative ear nothing

HOOKED.—THE FIRST RUSH.

can be more delightful than the infinite variety of the wild moorland sounds at this period of the year. The air is filled with the pipes and whistles of the wading tribe, and the varied intonations of the smaller songsters, with the croak of the snow-white Gulls floating overhead, and the "bec" of the Grouse-cock among the heather above—but, indeed, it is impossible to convey an idea of the variety of the bird-concert. The tamest of all birds is the Dunlin: a beautiful pair, with their rich ruddy backs and black breasts, run all round us, as we lie on the riverside at lunch, up and down the sloping bank almost within reach, and active and nimble as mice; they often wade breast-deep into the water to drink, and all the time keep "purring" in their usual careless manner.

The trout which have been taking freely these last few days, have suddenly changed this morning, and jump and roll over the flies without taking them. Later in the afternoon a severe thunderstorm, with heavy rain, came on. The effect of thunder on trout is curiously uncertain; but it undoubtedly causes them, in many cases, to cease feeding on the surface. On the 24th I was fishing one of the best streams in Northumberland: the water was in perfect order, brown as porter, and with just enough of flood. Yet the trout took extremely shyly—considering the season, and the perfect state of the water, quite unaccountably so; except that all day long there was a frequent, though very distant, rumble of thunder in the air, and this I suppose explained their unusual lethargy.

VANQUISHED.

The Pied Flycatcher should probably be considered rather as a local species than a very rare one in the north of England. Yet, though it has certain nesting haunts both in Durham and Northumberland, I never happened to see it in either county till the year 1885. The first occurrence was at Silksworth, co. Durham—a single adult male on May 7th (we had a slight covering of snow on the following day). The Migration Report of the British Association recorded an influx of these at the Spurn about May 5th, coinciding with the above observation. Later in the month I observed at least one pair evidently nesting in the beautiful woods below Wallington, on the banks of the Wansbeck. My brother afterwards discovered them nesting in another part of Northumberland and found a nest with six eggs on June 5th ("The Naturalist," 1886, p. 341.) This handsome little bird appears to be increasing in numbers, and was evidently very numerous in the year named.

These sudden, and more or less irregular incursions of what are otherwise scarce birds, are a well-known feature in ornithology. An example occurred at Silksworth in the spring of 1882. In that case it was the Grasshopper Warbler, a bird we had never before seen or heard there,—its most extraordinary note could hardly have escaped us had it ever occurred. We heard it for the first time on May 3rd, and for some time could hardly believe it was produced by a bird at all. The strange rattling flow of sibilant sound resembled rather the voice of a reptile or an insect—most of all a grasshopper, but we had no grasshoppers there. We afterwards observed several others in the neighbourhood, and on May 15th obtained a nest with four eggs. Except in that year (1882) I have never seen this bird at Silksworth, before or since.

May 20*th*.—The Wheatears are now laying: a nest in the heart of an old stone-dyke at Elsdon has four eggs. Two or three pairs of Whinchats also are breeding in the rough ferny banks at the Grasslees, but they are not a very common species on the moors.

May 21*st* (1887).—After several days of bitter cold and continuous northerly gales, the hills lay quite white with snow this morning—only a month from midsummer day!

May 24*th*.—The first young Grouse seen on the wing. They can hardly be ten days old, and no bigger than Sparrows; yet with the wind under their tiny wings, and the fall of the hill, one or two of them went at least 200 yards. This rapid development of the power of flight in the game-birds is a noteworthy feature, and the means by which Nature attains it are singularly elaborate. I am indebted to Mr. C. M. Adamson for permission to extract the following admirably careful observations on the subject from his "Scraps about Birds." Speaking of the Partridge, Mr. Adamson writes

(p. 7):—Nature, it would seem, has arranged that the little Partridge should be able to fly almost as soon as it leaves the egg; probably its tiny primaries and secondaries have begun to grow when it was inside the shell. Of course large feathers would be dreadfully in the way to such an atom, but it certainly soon gets such wings as enable it to fly. Well, its body begins to grow and get heavier, and, in consequence, it wants more power to lift it. But matters have been arranged for it without its taking any thought. Out comes one little feather on either side, and another, rather longer than it, begins to grow at once. So soon as it is partially grown, out comes the next feather on each side, and so soon as these are partially grown the same process goes on till the bird gets a new set of wing-feathers, proportionate to its increasing weight. No sooner have these all grown (or perhaps before they have) than another new set has in like manner begun to grow, and so a constant change is going on till the bird gets to a fair size. Then the final moult of both quills and tail takes place, at which period both old birds and young get their complete new set of feathers together. . . . The foregoing remarks relate only to the gamebirds. The case of Hawks, Eagles, Owls, Ducks, Geese, Wildfowl, Waders and all the small birds is quite different—all of these being entirely unable to fly till they get to their full size. In all these the quills come ONCE ONLY during their young state, and these quills have to serve the bird over the winter and until the general moult in the following summer.

May 31st.—Brood of nine young Wild Ducks on Redewater, newly hatched. I discovered them by noticing a small ripple coming from under the bank, and on looking over they all swam out, crossing the stream in a close flotilla. The old duck, finding her brood discovered, flapped up from some reeds in the river a few yards off,—the drake, of course, *non est.* Young Peewits are now flying in flocks, and Mr. Crawhall wrote me he saw yesterday, on the Waskerly reservoirs, in Weardale, a Black Tern (*Hydrochelidon nigra*), a rare bird in these northern latitudes.

SUMMER ON THE MOORS.

June is the leafy month elsewhere; and on the moors, where there are no trees, there is some equivalent for the absence of foliage, in the intense greenery of all vegetation. Even the heather is now green, and with ling and bracken, rush, grass, and sphagnum, all blend into a living carpet of greens of varied and vivid tones. In the vales, the golden bloom of the gorse is at its best about June 1st, and is followed up, a little later, by the broom and hawthorn. The ash-trees, always last, are hardly in full leaf till the middle of the month, but the spruce is now very beautiful—each dark green frond has an exquisite golden tip, and the cones their crimson tinge. The regularity with which, in the most adverse seasons, the hardier plants develop at their appointed time, is marvellous; as a Cheviot shepherd remarked to me, on Royal Oak Day, the cherry- and apple-blossoms in his little garden, "maun jest have come oot through the season of the year, for they've no had one right warm day to bring them oot this spring."

On a small lough, on the Scottish side of the Border, we enjoyed that day, what is always an intense pleasure to a naturalist—that is, making the acquaintance of a species which is new to him. Far away on the open water were several ducks, which the binocular showed to be certainly Pochards; it is a curious fact, that in all the years I have followed wildfowl, afloat and ashore, I had never before that day met with the Pochard in the north of England. In punt-gunning on the coast, we never meet with them, nor had we ever seen them before on the moorland loughs; yet the Pochard is described as a common species in many parts of England, and is, I believe, frequently obtained by punt-gunners as near as the Humber.

POCHARDS AND TUFTED DRAKE.

To face page 46.

Presently five of the ducks went ashore on a small flat island, where my brother and I crept within 50 yards of them, and at that short distance enjoyed a charming ornithological scene. There were four Pochards (three drakes and a duck), and one fine old Tufted-drake. A prettier picture as they sat on the low pebbly shore, some preening, others lying resting, and each form perfectly reflected in the still water, with a background of sedges and tall green flags, it would be difficult to conceive. There were at least a dozen Pochards (mostly adult drakes) on the water, and a still larger number of Tufted Ducks. From the date (Pochards lay about mid-May), there can be no doubt that these ducks were actually breeding there; hardly any of the ducks were visible, and the drakes most reluctant to leave the spot, circling round and round, high overhead, with rapid flight, and a harsh croak of a quack, more like the voice of a corby than a duck.

Tufted Ducks appear quite likely to breed in small numbers in the northern counties, and their having actually done so more than once is already on record. I have several times noticed pairs, and small companies quite late in the spring. But it should be remembered that the Tufted Duck (unlike the Pochard) is a late-nesting species, not laying till mid-June. Hence, their being merely seen at any particular locality during the earlier part of the summer, is in no way a proof that they will remain to breed there.

To the occurrences of what are called *rare birds*, I attach but little importance. There is no such thing as a rare bird, except in a relative sense. Go to its proper home, a few hundred miles north or south, and the supposed rarity is found as abundant as nature's balance of life will permit. All creatures seek out the zones of land or sea which best fulfil their requirements; when one wanders by chance or stress, a degree or two beyond those limits, it is regarded as a *rara avis*, and sentimentalists bewail the death of the poor belated straggler, as though, if it had been unmolested, the whole species would have extended their boundaries, or shifted their normal home. Personally, I would never harm such a bird myself, or give it other than a kindly reception ; but it is illogical to suppose that its death makes the slightest difference. In the vast majority of cases, a bird far removed from its natural sphere is destined to come to an untimely end, and even if it survived the hazardous experiment, would be most unlikely to wish to repeat it. Perhaps the most remarkable feature in connection with "rare birds," is that they should so often come across people who are able to recognize them. On July 24th, 1871, an Alpine Swift, 500 miles out of his latitude, was slowly hawking along the coast of Durham. Such a bird might easily spend a month there, without anything unusual being noticed; but on that particular day, Mr. Crawhall chanced to be walking along the cliffs, near Souter Point (with my Father and myself), and instantly recognized the species of the wanderer, not a score of which have ever been obtained in Great Britain. Since then, I have seen Alpine Swifts in dozens, at various places, and on Gibraltar have admired their superb dashing flight, making one dizzy as they hurl themselves over the 1,400 ft. mural precipice that fronts the Mediterranean.

Most remarkable of all wanderers is the Pallas's Sand-grouse, whose western irruption has attracted so much attention in the present year. The facts are too well known to need repetition here ; suffice it to say, that for the second time, a Central Asian species, whose home is in the desert steppes of Tartary, Turkestan, and Thibet, seized by a wholly inexplicable impulse, has left its far Eastern habitat, and followed the sun as far as land stretches to the westward. Crossing the Caspian and the Caucasus, these wholly exotic forms have spread themselves over Europe, from Archangel to Italy, and from Denmark to Donegal. The first great invasion was in 1863,* and a quarter of a century later—namely, in April 1888, we heard from Austrian sources of their second advance to the outposts of scientific observation. We were, therefore, prepared to hear of their advent at home ; yet I shall never forget the intense pleasure and surprise

* A minor invasion had previously taken place in the summer of 1859.

experienced on receiving (on May 31st) a newly-killed pair of these strangest and most lovely of birds. They had just been shot in North Northumberland, and were accompanied by a letter, asking what they were, and stating that a flock of sixty had arrived on May 6th, but that "as they were destroying his crop, the farmer had got leave to shoot them." The crops of these two contained a few grains of barley, and a quantity of what certainly looked very like turnip seed; however, on planting some of the latter, it proved to be the common field-runch, a useless noxious weed. Thus, so far from destroying his crop, the Sand-grouse were really assisting the farmer to clean his land.

PALLAS'S SAND-GROUSE (SYRRHAPTES PARADOXUS). MALE.
Nestling on the sand, as is their habit.

The great extent of waste sand-links in North Northumberland appeared to offer the wanderers at least as congenial a nesting-haunt as they were likely to find on British soil; and we did all we could to have the birds spared; but I fear this has been of no avail, for they had all left the spot by June 24th; though others were observed there later in the autumn, perhaps fresh arrivals.

By the middle of June the nesting season of the hardier moor-birds is nearly over. On the 3rd we found two Curlews' nests; one contained two rotten eggs and two young birds, grey-mottled, rather ungainly creatures, which, as their habit is, had left the nest on our approach, and lay squatted in the heather hard by. The other nest had one "dazed" egg—a local name, signifying that the embryo has perished after being partially developed; addled eggs are those which are unfertile from the first. The summer birds, however, are now at the height of their domestic cares. On June 10th, my brother found a nest of the Wood Wren (*Sylvia sibilatrix*) with 6 eggs, at the Blackburn linn, a rocky glen on Redewater, fringed with trees, and with a thick undergrowth of long heather and fern. Another pair were breeding in a straggling wood high up on the fells above East Neuk. This species, and the Willow-Wren, nest all over the moorlands, the latter in every wooded burn and glen as far up as the trees grow. The young Ring-Ouzels are now nearly full-feathered, yet on the 17th another nest contained five newly-laid eggs.

In 1877, on June 15th, I observed several Clouded Yellows (*Colias edusa*), and caught one—a butterfly I have never seen in the North of England before or since. Of the aristocracy of the insect world, the Emperor Moth (*Saturnia pavonia minor*) is a characteristic species on the moors, and its handsome green and red-spotted caterpillar may often be seen among the heather in the early part of the shooting season.

On the same day I found a nest of the Dunlin in a flowe on the highest part of Darden. The old bird fluttered off; her nest was on a grassy tussock, one of hundreds, each islanded in a labyrinth of black oozy peat-channels, and contained a single egg, almost hatching.

June 22nd.—Found a nest of the Twite among the heather on Lanshot hill. There were two eggs, and two small young. It is rather perplexing, at first sight, to find the white eggs of a *Linnet* where one only expects the sombre ones of a Titlark. My brother found another Twite's nest with four young, just fledging, on July 30th, on Elsdon Hillhead. At Silksworth the Tree-Pipits have eggs now, but I have rarely seen this species on the moors, and never chanced to find them nesting there. A Redpoll's nest on June 28th, in a thick osier-bed, had 4 eggs, and among the lining was a feather of a Kestrel.

The first young Black-game were observed on the wing on

July 10th, and the first young wild ducks on the 14th. The Mallard drakes have now entirely lost their bright green heads and handsome plumage, and are undergoing that remarkable "eclipse" which is peculiar to the genus and wholly inexplicable. Both Mallard and Teal drakes were observed apparently in full normal plumage up to the early days of June.

July 15th.—By this date the young Curlew, Plover, and other moor-bred wild birds are strong on the wing, and many are congregating into packs. Their southern migration is impending and will commence in earnest within a few days.

The month of July marks the conclusion of the summer period. Already among the feathered world there have begun to appear symptoms of autumnal conditions. As early as mid-June, the Starlings and Peewits are seen to be gathering into flocks; but in July the movement rapidly develops, and the signs of the time are plentiful and patent enough to those who are interested in reading them. Strange birds appear in strange situations. In the lowlands, the whistle of Curlew or Plover is heard amidst the unwonted environment of waving corn, or among enclosed fields of turnips or potatoes. From a farm pond, one perhaps springs a Dunlin or half a dozen Sandpipers; and at night strange bird-notes come down from the dark skies overhead. A "blackbird with a white breast" is perhaps reported by the gardener among the currant bushes. It is, of course, a Ring-Ouzel, and the small bird the cat has caught proves to be a young Wheatear. Poor fellow! he was just starting so blythely on his first (and last) voyage of discovery to the Mediterranean. On the seaside the Terns have broken up their nesting encampments, and spread themselves all along the coast, where the sand-eels and herring-syle are just now so abundant and so tempting. The Terns are immediately followed up by their arch-enemy, the pirate Skua, and almost every day one sees stray stragglers of the water-fowl and wading tribes—Whimbrels from Shetland, perhaps a little string of Grey Geese from Sutherland or the Hebrides, all in full cry, the first Godwits or Turnstones—the vanguard of the vast approaching hosts from the far north, flying from the arctic winter.

It is unnecessary to recapitulate in detail all the manifold sights and signs which serve to demonstrate the feature of the season. The bird world is on the move. The nesting season is over; the cares of the spring and summer are past; and the universal southward movement towards winter quarters has commenced. It is conspicuous enough in July, but attains a far greater development in August, and approaches its climax when the Swallows are seen congregating on the trees in September.

A SUMMER RAMBLE ON CHEVIOT.

On a magnificent summer's day—one of the finest our temperate zone is capable of producing—when a blazing hot sun is tempered with a cool northerly breeze, we set out for an ascent of Cheviot. An hour's walk over the outlying spurs brought us to the foot of the Caldgates valley, one of the most charming moorland glens on the Borders—certainly the prettiest on the granite formation—and far more variedly luxuriant than the valley of the College, the alternative route by which Cheviot is approached from the other side.

One almost wonders why the natural beauties of our own land are nowadays so neglected by travellers. Without at all depreciating the grandeur of other countries or the charm of foreign travel, it does seem regrettable that our fine wild scenery at home should be ignored and all but unknown. The route up the Caldgates glen leads through three or four miles of lovely moorland scenery, almost Swiss in character; the track following the course of the burn, a rocky, splashing streamlet alive with trout, and fringed with patches of gorse and straggled belts of natural wood—birch, oak, alder, and rowan. The bloom of the hawthorn was perfect—each bush a canopy of spotless white as pure as new-fallen snow; the mossy banks and braes glowed with the purple of the wild thyme and heath, and were alive with grasshoppers and small chestnut-winged butterflies. Above the woods, on either side, the heathery hills rose steep and rugged, great naked rocks standing out here and there as abrupt as ruined castles. The air resounded with the notes of Willow-Wren, Ring-Ouzel, Wheatear, and Sandpiper, and overhead floated scores of white Gulls from Pallinsburn.

The charm of the moors lies in their pristine beauty of creation, unaltered and unalterable by man. His presence, indeed, is hardly perceptible; the only specimen of our race seen all day was an old man with a vast bushy beard and a pack on his back, who was resting in the shade of the trees and trying to light his pipe with a burning glass. He was "jest seeking a wee bit pickle o' 'oo'" (wool), and had walked "from *Scotch* Belford, no' that awfu' far," though by the map I see it is a dozen miles or more. Poor old soul! he reckoned a pound of wool, worth sixpence, a fair good day's pick, and spent his life wandering about these wild hills gathering stray scraps of wool and depending on the charity of the shepherds for chance accommodation. Yet he was not a tramp: that is quite a different species, and one that is remarkably abundant all over the Borders. At the head of the glen lies Langlee-ford, a lonely farmstead and the last house in England, beautifully ensconced among pine-woods—a protection from the snow-blasts that in winter sweep down from Cheviot. To-day, however, the heat is tropical, but for the light breeze that comes laden with the delicious fragrance of the pines and the hawthorn, of the rowan and woodbine, and a score of nature's exquisite perfumes.

From Langlee-ford we "take the hill," and the climb commences in earnest. At first the ascent is over ordinary moorland, with bracken-beds, now in their beautiful emerald green stage. From the heather close by spring three or four cheeping half-grown fledgelings. They are young Grouse; and at the same moment there is a flutter and scuffle a few yards away, as their anxious mother flaps along the ground as though winged and disabled. How admirably she diverts one's attention at the very instant her brood need an opportunity to escape unseen! Not till they are all in safety do the old Grouse take wing—the old cock all the time crouching within a few yards. Grouse are noble parents—very different to their cousin, the Blackcock, who after the vernal courting, never again looks near his numerous wives and families.

Leaving the gaunt cone of Hedgehope on the left, the flat summit of Cheviot presently comes in view, still far above. Gradually, as we ascend, the heather grows less and less luxuriantly, becoming scant and dwarfed, and mixed with the golden leaves of the bleaberry-ling, the whortle-berry, and the creeping heath. For the last few hundred feet the vegetation is so stunted as to resemble a great soft mossy carpet, as easy to the tread as those of Turkey, though perhaps not so smooth, since strewn broadcast on it lie patches of the dark grey rocks—porphyry, dolerite, and granite. The actual summit is a broad flat plateau, perhaps half a mile in extent. Bleak and wild-looking, the plateau is only half-clothed with coarse bent and cotton-grass, interspersed with barren mosshags, oozy peat-flats, and ravines. The small white flowers of the cloud-berry (*Rubus camimorus*), a plant which only flourishes at altitudes of some 2,000 feet, were a relief to the monotony of barrenness, together with tufts of Lycopodium and the trailing shoots of the crowberry. The Alpine *Cornus Suecica* also grows at one spot here—a very rare British plant, only found on Cheviot and on one other of the northern fells. The only birds seen on the summit (2,676 feet) were a Grouse or two—none nest so high—a few Golden Plovers, and—a charming sight—quite a small colony of Dunlins. There were five or six pairs of this graceful little wader, all breeding together among some moory tussocks, and extremely tame, perching within a dozen yards. We sat and watched them for some time with the binocular—pretty little chestnut-striped birds, with a black patch on the breast.

On a bright, clear day the view from Cheviot amply repays the labour of the climb up. The eye ranges over a panorama of wild mountain land. Looking northward across the fertile vale of Tweed, with glimpses here and there of its silver thread, the horizon is bounded by interminable Lammermuirs. The triple crests of the Eildons, above Melrose, are prominent objects to the west, while all the succession of rolling fell ranges along the Border are clearly distinguishable. In the far-away distance the steam of a train in the "Waverley route" seems incongruous, so we turn southward. Here, too, there are hills—nothing but hills. Kelso Cleugh and the Windy Gyle, the broad contour of Shillmoor, and, close at hand, the rival peak of Hedgehope, whose smooth green slopes are furrowed and excoriated with black fissures and peat-cracks, like the pencillings on a Bunting's egg.

The bold black outline of the Simonside range limits the discernible view southward, though beyond it are visible blue vistas stretching away beyond the Tyne. Eastward, nearly the whole seaboard of Northumberland lies in view— Holy Island, with its white sands shining in the sunlight against the blue sea; the ancient Border fortresses of Bamborough, Dunstanborough, and the Lindisfarne; the wooded heights of Chillingham and the fatal field of Flodden; farther away, the Farnes and Coquet Island, dimly seen through a slight sea-haze.

Few spots on the British coast are more interesting both physically and historically than this corner of North Northumberland. The singular rock formation of the Farnes —the sandy wastes and dreary mud-flats of Holy Island, covered and uncovered twice every day by the sea—

> "The tide did now its flood mark gain,
> And girdled in the saint's domain;
> For, with the ebb and flow, its style
> Varies from continent to isle;
> Dry shod, o'er sands, twice every day,
> The pilgrim to the shrine finds way;
> Twice every day the waves efface
> Of staves and sandalled feet the trace."
>
> *Marmion.*

Here, twelve centuries ago, St. Cuthbert established the cradle of Northern Christianity, choosing the sequestered Lindisfarne for his island home, hard by where, to-day as then, the surf breaks white on its basalt barriers, and the Cormorants go to fish for codling in the swirls of the northern sea.

In all, five counties lie stretched out before one, a wild landscape, the scenes of bloody days and of ages of human strife. Thence we ramble along the crest to a rocky peak where Arkhope Cairn marks the actual boundary. Here England and Scotland are divided by a mountain gorge as wild and bold as can be seen in the three kingdoms—its sheer, smooth slopes descend some 1,500 ft. on either side, only broken by abrupt black crags and ridges of porphyry. The College burn springs from the bowels of the abyss, and its wild, romantic valley lies full in view far below, terraces of ancient moraines and hanging woods impending the stream. Overhead croak a pair of Ravens, mere dots in the azure height, and an old Blackcock speeds away, disturbed by a boulder we send leaping and crashing downward into the depths below.

We have over a dozen miles to tramp home, and the shadows lengthen. The steep slopes we traverse are now

"WHERE ENGLAND AND SCOTLAND DIVIDE.

all orange and gold, with the bright-tipped leaves of the bleaberry aglow in the evening sunlight. This quasi-Alpine shrub here entirely displaces the heather, and we noticed the bright yellow flowers of the tormentilla, like a small buttercup, and a few tufts of saxifrage. Perhaps from an ornithologist's point of view the most interesting event of the day was the sight of a Peregrine Falcon: this noble bird had evidently been resting on a pinnacle of some bold crags which lay in our course, and dashed out from below with a loud, oft-repeated scream. In a few seconds he was up in the clouds, and then, for a time, wheeled and circled overhead, while we enjoyed the exhibition of superb grace and the powerful ease of his flight—a sight worth walking twenty miles any day to see. In due course we struck the head of a small hill-burn which led towards our destination, and for some six miles followed its stream along heath-clad moors, where the Curlew whistled and Plover piped; by tumbling cascades, where the rowan scented the breeze and the foxglove grew in rocky crannies; by dark pools, where the trout were splashing and playing: a lovely walk in the cool twilight ends a delightful day.

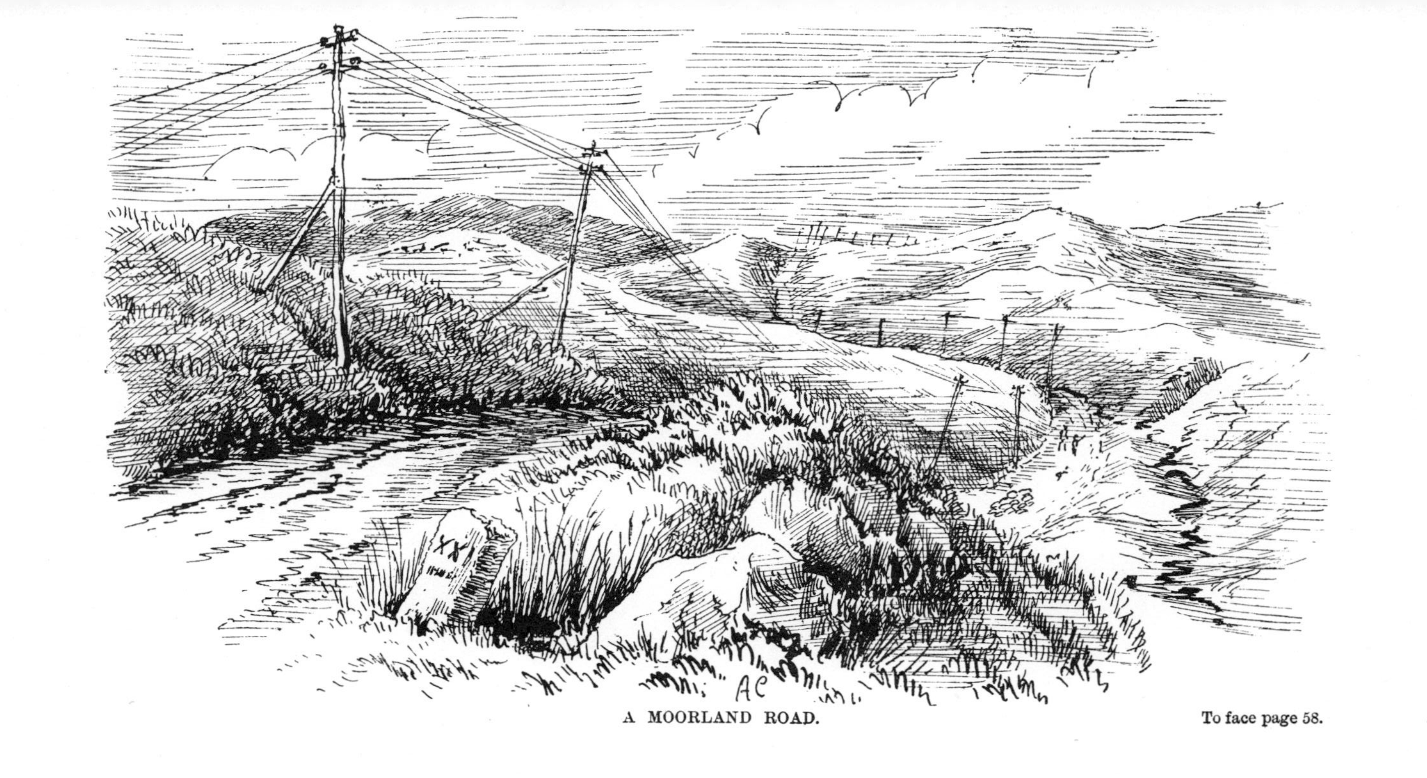

A MOORLAND ROAD.

To face page 58.

SOME RECOLLECTIONS OF THE "TWELFTH."

WHAT is the meaning of the word "wild" as applied to Grouse in August? It is often difficult to understand what degree of wildness is meant, especially when, as one often reads, the expression is appended to a report of perhaps several hundred birds having been shot. Perhaps it is merely a *façon de parler*, a form of words to magnify the exploits or gratify the vanity of the shooters. Obviously, when grouse are really wild, they cannot be killed in hundreds over dogs or otherwise (except by driving). By comparison with the numbers of people who flock to the moors in August, those who follow the sport of grouse-shooting throughout the season are very few; but it is the latter alone who really know what "wildness" means. Then, in late autumn and winter, it really means that in the contest "Vir *v.*Tetrao," the former is nowhere, is outmatched. In keenness of sight and power of locomotion, in ceaseless vigilance, he must acknowledge his inferiority. On the open fell in November man is comparatively powerless; it is only by his ability to work out schemes of stealth or strategy that he can bring himself into the same acre with his noble quarry.

Yet in the August reports it has become almost a set phrase, "birds wild and strong on the wing," a common affix being "scent very bad." Now, this (the former) must very often simply mean that the young grouse are normally well-grown, and rise boldly at perhaps thirty or forty yards, instead of "cheepers," which can be poked up from under a dog's nose. Young grouse hatched at an average date, say mid-May, are by the "Twelfth" full three months old, and in the ordinary course of nature are nearly full-grown, and have their powers largely developed. Such birds have cast the soft spotted quills of their nestling plumage, and acquired the strong black primaries of winter, together with a large proportion of the adult dark-red plumage. Only a streak of the yellow-barred nest plumage remains along the centre of the breast, dividing the newly-acquired dark feathers on either flank. At that stage, in a species so bold and intractable as *Tetrao scoticus*, it is not reasonable to expect close point-shots, at any rate *at the first rise.* They must be followed again and again, out-manœuvred, broken, and "dominated."

But this leads me to the second part of my text—the bad scent question; and here may I say that these remarks are not addressed to "masters of the art," but are written with a desire to be of some assistance, however slight, to the many who have not opportunities to master the subtleties and the *minutiæ* of successful grouse-shooting, to some of whom the "Twelfth" oft-times brings only vexation and the chagrin of fallen pride. Assuredly walking about a moor with a gun does not constitute grouse-shooting. Well, then, bad scent —I write from frequent observation—often means *bad dogs.* Not but that there are days when scent *is* bad, sometimes almost *nil*; but, in a general way, scent is best to the dogs which best understand their business, and have been taught how to avail themselves of what scent there is, and *vice versâ.* It is a fairly safe rule to lay down that a good man will have a good dcg, simply because they appreciate one another; and in proportion to the extent of their mutual confidence, the directing power of the mind of one is able to bring out and develop the instinctive faculties of the other. The writer claims to be no authority on dogs, quite the reverse; but, in shooting regularly over them, one cannot but see that there are dogs and dogs. There is one class of dog—very numerous—which appears to regard a hunt on the moors as an institution especially arranged for their sole and particular delectation. They travel far and fast; they persist in hunting one hillside while their owner is helplessly endeavouring to work another, perhaps hundreds of yards away. For his wishes his canine assistants (?) care nothing. The keeper, with stentorian lungs and an ear-piercing whistle, which would do eminent service as a "siren" on the outlying rocks of some fog-enveloped coast, endeavours to induce them to hunt the same grounds as his master, but in vain. These dogs are often what is called "broken,"—that is, they have had hammered into them the mechanical lesson (if within earshot) to go down to hand, and also to back each other's points. Usually this accomplishment is purely mechanical, and one has to laugh at the absurd spectacle of one of these valuable auxiliaries to human nasal deficiencies steadily backing another which has stopped to get a drink, or for any other purpose. Then, when they find birds, it is usually at least dubious if the latter will lie till the shooter travels, say, a quarter of a mile to his point, even if the mechanical training is good enough to insure the dogs not springing the game by jealously drawing in too near them before the gunner gets up. With such dogs, a hardworking man on a wide beat may get a fair bag of grouse in a long day; but he would perhaps get almost as many without running a dog at all.

The opposite type of dog is that which appears to realize that its object is to assist its master to kill game; which persistently and closely quarters the ground right in front of him, attentive to every signal by hand to pay special attention to likely bits; which seldom hunts far out of sight, even when in hollows, or places where, by reason of the lie of the ground, one can only see a short distance; which takes advantage of the wind, so that one can practically hunt in any direction almost regardless of its "airt"—in short, which utilizes to the full every canine instinct and faculty in co-operation with its master's directing capacity. Nothing to my eye is more ridiculous or less workmanlike than to see a man spending nearly half his time walking away to leeward so as "to give his dogs the wind" on the return cast.

As already mentioned, it is well known that the wildest grouse in August can be dominated by persistent following up. But it needs useful dogs to find them again and again, for, after a few rises and a long shot or two, they become demoralized, and will then hardly rise at all. This is best illustrated if one happens to be at the spot where a covey stop after being shot at. One hears the shot, and, sitting down, presently sees a straggling line of grouse top the ridge and dip down the sloping glen. Suddenly one of them stops —appears to dive headlong into a patch of long heather within ten yards of the ridge. The rest hurry on, closely hugging the heather, and at the bottom three or four more tumble themselves headlong into the covert, all scattered, and, in their fright, apparently utterly careless as to how they alight. Fifty yards further and down plumps another, then another, till all have gone to ground. One sits still and waits patiently, knowing friend "Siren" will be following up. Soon he appears; walks right past number one, and numbers two to four, before his dogs have appeared on the ridge. Then they rush upon the scene; take a few wide gallops—no systematic search, *sine quâ nihil*—perhaps never within a hundred yards of the lowest birds. "Siren" remarks that they "must have gone on!" and himself does likewise. Then one can pick up in half an hour three or four brace of fine young grouse to close points, on the ground where friend "Siren" has only had a long shot at the old cock.

Now, this is not grouse-shooting; but if "Siren" happens to report his bag, he will probably add that the "grouse are wild and strong on the wing," &c. No one can fairly dispute his ingenuousness, for he is wholly unconscious of having again and again walked right past them; and, if he also blames the scent, he is quite innocent of his own ignorance in handling his dogs.

One remark as to the habits of grouse. It may seem impossible to say anything new on so trite a subject; yet Wellington was, perhaps, right when he said that nothing is impossible; for there exists a very prevalent and erroneous belief that grouse feed twice a day—morning and evening. Mention the subject among a party of grouse-shooters on the evening of the 11th, and prove whether this statement is correct or not. Grouse never feed till evening, but the error alluded to probably rests on the shadow of authority which is lent to it by the mistaken statements in print by several writers on the subject. No doubt, also, the grouse do give some grounds for the supposition by their habit of "flighting" at daybreak, and by being generally found during the early hours of the day on short "feeding heather." But, as already stated, grouse never feed till evening. I speak only of heather-fed birds, having had no experience of those which (by the proximity of oat stubbles) have been "educated" up to a corn diet. That this is the case can easily enough be proved. Open the crops of a dozen grouse, say at 8 or 10 A.M.; they will be found empty, except a few stray heather shoots or rush seeds. These are apparently picked up in pure carelessness, or for amusement. Perhaps there will also be found a few bits of gravel, taken to aid digestion. But every one knows that the crops of grouse killed towards dusk are choke-full of heather shoots; an old cock will contain a breakfast cupful. I mention this point in order to show that any "strategic operations" based on the assumption that the grouse will be on the feed in the early morning, are (*q.v.*) undertaken erroneously, and also to show the advisability of holding in reserve till evening a fair proportion of "going power," human and canine. Then, in the closing hours of the day, the reserve power can be utilized most effectively. The birds being scattered on the feed are easier to find, lie closer, and are more apt to rise separately. Thus each covey will, perhaps, yield several fair shots, and between 5 and 7 P.M. a moderate bag may be converted into a heavy one by the sportsman who knows how to abide his time.

To experienced and skilful grouse-shooters these few random notes will, no doubt, appear trite and crude enough; but all are not experienced, and it is a prevalent mistake to regard dogs as mechanical automata, and grouse-shooting as merely an affair of so many hours' walking on the heather. Possibly the perusal of these few hints (should they do me the honour to digest them) may be of some advantage to the rising generation, of whom every succeeding year brings forward a fresh contingent, to enter for their first "Twelfth."

Post Scriptum.—In the Badminton Library (" Shooting —Moor and Marsh," p. 3) exception is taken to the above remarks on the feeding habits of grouse. The author, one of the first sportsmen of the day, questions rather than disputes their accuracy, holding that the evidence adduced does not amount to proof. Possibly, to that extent, his contention is correct; but how singular it is that one of the everyday habits of a game-bird, which, for more years than I can tell, has been studied and pursued annually by thousands of sportsmen, should still, apparently, be open to question?

STRAY NOTES ON GROUSE AND THE GROUSE-DISEASE.

So many able scientists and experienced observers have discussed the subject of grouse-disease, and promulgated theories (many of them at variance, and none apparently conclusive), that the writer hesitates to express any opinion as to its exact origin. One circumstance, however, appears to be invariable, and to be the inevitable precursor of disease, viz., a heavy stock of grouse. I have never myself known of an irruption of disease except after more or less plentiful periods, though it may occur. Certainly it is the inevitable consequence of an undue plenty; the cycles are almost regular—plenty and scarcity in constant sequence, though the periods of each vary.

Different areas of moorland vary greatly in their power of sustaining a head of grouse. Thus, on a hill range in Perthshire or Aberdeen, every acre may accommodate (say) two or three grouse, while in the lowlands, or in Northumberland, one grouse to three or four acres may be a full stock. But each moorland area has its fixed capacity, and, whatever the local maximum may be, when it is exceeded disease is the inevitable result.

This variation in grouse-productive or sustaining power is exemplified in the north of England by the immense head of game which in some years is attainable in North Yorkshire and the adjoining moors of Durham. The deep peat deposits and rich heather of Teesdale in Yorkshire, and of the Weardale hills in the adjacent county of Durham, are vastly more prolific of grouse than are the more alluvial moors of Northumberland, extending northwards along and across the Scottish border. In these latter districts the heather is of less luxuriant growth, and alternates with stretches of white grass, rushes, and bracken. Moorlands of this character, though eminently suitable for blackgame—of which they are, *par excellence*, the stronghold—are inferior in grouse-producing power; where both species of game are found together, the power of man to increase abnormally the head of grouse is limited. It is not till the Scottish Highlands are reached that we find repeated on the solid heather of Perth and Aberdeen the phenomenal fecundity of Wemmergill and Blubberhouses.

To resume: man, beyond doubt, is the primary cause of grouse-disease in his tampering with Nature's balance of life, and with the natural conditions on the moorlands. Nature designed various checks upon the undue fecundity of the *Tetraonidæ*. She formed the Peregrine, the Harrier, and the Merlins specially to hunt the moors. Man determined to have all the hunting himself, and has removed Nature's safeguards. Doubtless a century or two ago all the above-named birds of prey abounded on the heathery uplands, and day by day examined every acre of fell and flowe, picking out both the superabundance of healthy birds, and the sickly if ever the symptoms of disease appeared. Thus the disease was practically unknown till some time after the commencement of the present century.

But now we have changed all that. Game preservation and vermin trapping have created a new order of things. The Harrier and Falcon have gone; the hill-fox and weasel are held in check. Thus the stock of Grouse is vastly increased, and is maintained ever close along the margin of the dividing line beyond which Nature has decreed it shall not go. When that line is passed, she re-asserts her supremacy, repels our interference, and disease sweeps bare the heath-clad hills.

The extent to which the system of heather-burning is now carried, is unquestionably another factor in the promotion of disease. In the fierce competition of the age, the heather, like the grouse, is, so to speak, "forced"; an unnatural quantity is demanded from the hills, and an artificial state of affairs created, both as regards the crop of heather and of grouse, which are but "animated heather." This straining of Nature's gifts to the utmost, must necessarily tend to cause deterioration in the quality of both, rendering them more liable to injurious influences, or less able to resist their attacks.

The grouse-disease appears to be divisible into two separate and distinct types. One, the most common, is the lingering form, which is slow in its operation, gradually reducing its victims to mere skeletons, when they die apparently of emaciation. The symptoms of this malady are, first, in the grouse affected seeking lower ground, especially along burn-sides and wet places, sometimes right down in the valleys where sound birds are never seen; and, secondly, in the change of plumage, which loses its fine glossy sheen and fades to a dull dingy hue, most unhealthy-looking to a practised eye. Their legs and feet at the same time lose the feathery "stockings," and become bare and draggled.

The other type of disease is much more subtle, more rapid in operation, and less easy to foresee; indeed, its approach is often hardly perceptible—it comes "as a thief in the night." A few years ago (1884) we had in Northumberland an eruption of this sudden form of disease, of which I propose to describe the principal features. We had heard, during the spring, intermittent reports of the appearance of disease in various quarters, and particularly on certain specified moors. After the abundant season of 1883, grouse-shooters were nervously apprehensive of what might occur; but up to the middle of June their fears were certainly baseless, and (at least on the writer's ground) there was no reason to suspect the approach of disease. In the course of several visits to the moors during that spring, I could detect no signs of anything really serious—nothing worse than an old bird or two "found dead." Early in June I examined the ground carefully. Nothing could well appear more favourable. The young broods had hatched out in great numbers. Many could already (on June 1) fly two hundred yards or more. The majority, however, were still "in down" in various stages. Only a very few nests contained eggs. These I noticed were rather less richly coloured than average grouse-eggs; but they nearly all hatched out subsequently, except one or two, which were almost colourless. In the middle of the month (June) I had a very favourable report from the keeper. "I have seen a lot of broods to-day," he wrote, "one with eleven, a grand lot. I think the disease has now quite stopped, as I have seen none new-dead lately." So matters ran on for a whole month. But the line had been passed; the inevitable result was bound to follow; and at the end of July, literally at the eleventh hour, broke out a disease whose deadly virulence devastated the hills, and in *less than a fortnight* the stock of grouse was decimated.

Here is the report of August 5: "I am sorry to say the prospects for the 12th are very bad indeed. The 'black ground' by all means worst, as that was sure to be when disease comes; mixed ground is always best off. I was out yesterday all over the best ground with the dogs. I found the young birds dying—great big good birds. I opened some, and it is the real disease—their livers affected. I am certain I could not have shot five brace at the outside, and hunting all the best of the ground."

The outward symptoms of the disease in this sudden or acute form were not very easy to recognize when shooting in August. The physical condition of the grouse, their state of plumpness or leanness, was hardly any criterion. Many birds which were undoubtedly affected still retained their full plump breast and thighs, and the proportion of emaciated, fleshless grouse, with protruding breastbone and stocking-less legs, was insignificantly small in relation to the extent and virulence of the disease. Nor was their plumage in this case any more reliable an index; it is, moreover, always worn and dull at this season. Keepers, however, in years of disease usually forget this (or do not know it), and ascribe the washed-out appearance of old birds exclusively to the disease. From the above and other reasons, I infer that this particular form of grouse-disease is no lingering illness. It cuts its victims down sharply and suddenly, before they have had time to lose their plumage or condition. For a few days after its attack, the grouse shows little or no visible change externally. When emaciation sets in, death follows almost immediately, within a few hours.

The best indication of the presence of disease in this form appears to be the manners of the old birds and their proportion to the young. In the year in question (1884) I was at first a little deceived by appearances on the Twelfth. Expecting to find the hills almost bare, we only took out half an ordinary supply of cartridges. Consequently, when, between three and four o'clock, having then fired my last cartridge, and with some of the best hours of the day still to come, I sat down with twelve brace of Grouse, a Teal and a couple of Golden Plover—a fair bag on that ground in any year—I felt inclined to anathematize the exaggerated reports of disease. Exaggerated they certainly seemed, for birds had the appearance of being tolerably numerous, and signs of disease but few. I felt sure, had cartridges lasted, of getting twenty brace. But the next time we went over the same ground, the true state of the case became conspicuously apparent. With plenty of cartridges and a long day, it was only by hard work I managed to get together five and a half brace, and they all old birds! The fact is, this form of disease at first helps the gun. Old birds (broodless) which would normally spin away, a dozen at a time, at two or three hundred yards, now rise singly and just within shot. But as soon as these are killed, it is all over. Ichabod! There are none to take their places.

So matters remained throughout August and September. Of young broods there were simply none, and it was but labour lost to hunt for them: the young had evidently been the first to succumb. But the disease appeared to have been local, and had not perhaps affected any very great extent of ground, for in October, on the general movement or "re-distribution" of grouse which annually occurs in that month, our stock quickly rose to a normal level, and continued so for the rest of the season.

As the subject of "vermin" has been alluded to as one of the factors in the production of disease, the following statistics, showing the results of their depredations on moorland game, may perhaps be appropriate. The ground vermin—fox, stoat, and weasel—are by far the most destructive, and their reduction by trapping is essential to keeping up a fair head of game. The hill-fox, which has cubs at the time when the Grouse are sitting, is the most deadly enemy to them, taking the hen birds off their eggs; but the stoat and weasel, being far more numerous, are perhaps almost equally destructive in the aggregate. If these three pests (I refer, of course, to non-hunting districts), and the Corbies, or Carrion Crows, are kept down, probably a few Peregrines would not do a very perceptible amount of damage, and the small Hawks even less. The following figures, showing the quantity of game killed on the same ground during two equal periods—(1) without trapping at all, and (2) with regular trapping all the year round—are sufficiently eloquent on the subject:

Game killed.	(1) PERIOD. Without trapping at all.	(2) EQUAL PERIOD. Trapping regularly.
Grouse	1109	2125
Blackgame	308	518
Partridge	89	201
Pheasants	2	6
Hares	13	29
Snipes	122	228
Plovers	105	130
Ducks	13	33

The grouse-disease is unquestionably the price we have to pay for maintaining the stock of moor game at a much higher level than Nature ever intended; but, on the whole, we are no doubt very greatly the gainers. For one season of bad disease, we have perhaps five or six of an artificial or extra-natural abundance.

In corroboration of the above deductions, it may be interesting, in winding up these rambling notes on a bird which possesses such importance to sportsmen, to mention that in Norway, where the closely allied Willow-Grouse (*Lagopus subalpina*) is extremely abundant, grouse-disease is practically unknown. The Norwegian species is nowhere more numerous than in Lapland, where, according to Wheelwright (" The Old Bushman " of the *Field*), no signs of disease have ever been detected ; and the same remark, I believe, would apply to vast tracts of fell land throughout Norway, where man's disturbing influence has never made itself felt, and where the grouse are left to fight unaided their own battles for existence.

BIRD-LIFE ON THE MOORS IN AUGUST.

On reading year after year in the August numbers of the *Field* the results of the annual campaign on our northern moors, and considering the immense extent of wild country which, after eight or ten months of unbroken peace, is simultaneously invaded and searched out by man and dog, it always strikes me as remarkable how few wild creatures, save the game, come in the way of the hosts of guns. From scores of moors, forests, and fells arrives the almost unvarying record—grouse, nothing but grouse. Some, perhaps, who only see the moors during the season of purple heather, may conclude that the wild hills are rather deficient in variety of bird-life. The " Twelfth," in point of fact, falls at what happens to be ornithologically one of the least interesting periods of the year—between the departure of most of the summer birds and the arrival of the winter ones. Most of the former, whose beauty tends so much to enliven the heathery solitudes during spring and summer, have reared their young and departed. Some have disappeared entirely ; for example, I have never observed a Redshank or common Sandpiper—both of which are numerous in spring—remaining inland so late as the "Twelfth." The Black-headed Gulls too are gone, though occasionally an exceptionally late straggler may be met with. Such are almost invariably the brown-mottled young of the year, probably late hatched. The Dunlin I have only once come across in the shooting season —a newly-fledged young bird, which I shot on August 15th in the backward season of 1879.

Of the moor-bred fowl, amongst the most conspicuous in spring are the Curlew and Golden Plover. In both these species a great movement has taken place between the middle of July and August 12. The Curlews have almost entirely disappeared from the fells by the latter date, but a few still remain in the moorland districts, and in mild wet seasons continue to frequent the rushy fields and low-lying "haughs" till much later in the season. A few late-hatched young Curlews may also be found about the places where they were bred. These are often quite tame, and, lying close in heather or long rushes, I have killed them to "point-shots," sometimes after a "rode" of fifty or a hundred yards before the dogs—a remarkable occurrence when we remember the wild nature of the old Curlew.

As regards the Golden Plover, it is my opinion that the great majority of the local-bred birds have departed for the south before the "Twelfth." Their southern migration commences at the end of July, and very few of the birds actually bred on any given moor remain on it till the middle of August. On the high "black ground" where they breed, we seldom see many in August (except sometimes in wet weather)—only a solitary ragged old bird with marbled breast, or a late hatched young one still downy on the neck, which pipe about by themselves, and generally manage to get shot. The Golden Plovers being found more or less permanently (*i.e.*, at all seasons) on the moors, their migrations are less easy to trace than those of birds which depart entirely; but it appears probable that those found on the Border moors in August are the produce of the Scottish Highlands, Shetland, &c. These pass along southwards in a continuous but irregular stream, commencing about the period of the departure of the local-bred Plovers, say end of July, and continuing almost till the arrival of the great flights which come from Norway, Lapland, and Siberia in September and October.

Snipe, too, are in the same category—*i.e.*, of birds which are migratory as a *species*, but which, as *individuals*, are found here at all seasons. Many of those shot on the Border moors in August are undoubtedly the native-bred birds, being in all stages of adolescence from the down upwards. It is not unusual to find young Snipes which can only fly thirty or forty yards, and on August 12, a few years ago, I found a whole brood just emerged from the shell. This was of course an exceptionally late case, the normal period at which Snipe begin to lay being the middle of April. Precisely the same remarks are applicable to the movements of Snipe during August and September, as I have already made in reference to Golden Plover. In September we certainly have birds on passage, arriving irregularly, yet their quills appear insufficiently developed to have enabled them to cross the North Sea. Hence it is probable that these are Scotch-bred birds on their passage south, the foreign migrants not arriving till October. There usually occurs a perceptible interval between the departure of the last of the British-bred wild birds, and the arrival of the foreign contingent. The interval varies in duration in different years, and is observable both with Snipe and Golden Plovers. Very hot, dry seasons are of course unfavourable for all these birds, which disappear from the inland moors entirely; but in such seasons I have sometimes noticed a corresponding increase in the number of Peewits.

Of course all grouse-shooters meet with the birds above-mentioned, together with the rest of the regular moor-bred wildfowl, such as Mallard, Teal, &c.; but how seldom one hears of the occurrence of any really uncommon species! Considering all the circumstances, this appears rather remarkable. For example, one might expect to hear of some of the less common ducks being obtained, or of some of those ducks, waders, &c., whose usual summer habitat is in northern latitudes, but which have been stated to remain occasionally to breed in the wilder and more remote parts of the British Islands. According to some of the highest authorities on ornithology, there are several species of our winter wildfowl of which a few pairs do remain to breed with us. If that was generally the case, the northern moorlands would certainly offer such birds the most attractive and congenial home; some of them, consequently, might reasonably be expected to fall in the way of the invading army of gunners in August. That they do not do so is strong negative evidence against the supposition in question. Personally, I have never shared the opinion, either that these wildfowl breed in England, or that by legislation they may be induced to do so; with the utmost deference to those who think otherwise, I fear it is only "the wish that is father to the thought." England is not the summer home of the great masses of migratory wildfowl. Neither Wild Geese, Pintail, Wigeon, Golden-eye, Scaup, Long-tailed Duck, Goosander, Merganser, or either of the Scoters have ever bred here, and they never will. No doubt a few pairs of Shovellers, Pochard, and Tufted Ducks nest with us more or less regularly; but their numbers are wholly insignificant. What do they amount to? Possibly a hundredth of one per cent. on the aggregate numbers of their species. In the north of Scotland and the outer islands, it is true, a good many Grey-lag Geese, Wigeon, and Mergansers remain to breed; but their numbers, after all, are the veriest trifle, and the whole Scottish supply a mere drop in the bucket. In the south of Spain, I have seen more Grey-lags and Wigeon in a single day than Scotland would produce in a score of years, and as many Shovellers as would hatch there in a century.

There have not been wanting, nevertheless, ornithologists who have undertaken to show that nearly all sorts of wildfowl breed in the British Islands; and they have succeeded, too, after a fashion. Facts, however, remain unchanged despite the desires of theorists and the sophistries of those who are, perhaps unconsciously, too apt to bend to the current of any *popularis auræ*, and it is upon such slender bases that our wildfowl legislation rests. Such arguments are hardly ingenuous, inducing beliefs that are contrary to known facts. The province of science is surely to teach, to make clear, not to mistify by confusing exceptions with rules, rules with exceptions. The legislation in question is a sad story of ignorance, prejudice, and bungling, but little in harmony with the broad lights of ornithological knowledge of the present day.

I cannot help thinking that into this question a degree of credulity has been imported, which in all other cases is wholly discountenanced ; and evidence is accepted which an unbiassed mind would at once dismiss as valueless. Thus a fictitious importance has been attached to chance appearances in summer of any of the Arctic-breeding fowl—ignoring the wide difference between the Arctic calendar and our own. Here, April is the nesting-season ; there, July. Then there are several species of the duck-tribe which certainly do not breed at all in their first year : and the young (immature) birds of such species sometimes remain here throughout the summer. We have Scoters, for example, on the north-east coast all the year round, but it would be absurd to conclude therefrom that the Scoter will ever breed here. It should be borne in mind that no amount of probabilities establish a fact ; but, on the other hand, it may fairly be pointed out that the young of many of the ducks, &c., differ so materially from the adults that, even if obtained, they might not be recognized, or be mistaken by many sportsmen for the commoner kinds.

Of the duck-tribe, the only kinds I can myself record during August, excepting of course the Common Mallard and Teal, are single occurrences of the Tufted Duck and the Shoveller. My brother shot a female Shoveller on Aug. 12, 1877, on a small rushy lough near the coast ; and, though I did not actually obtain the Tufted Duck (which escaped, broken-winged, by diving), have no doubt as to the species. Both these ducks are known to breed in Northumberland.

Several of the strictly summer birds remain in the moorland districts till the end of August, or later. The Cuckoos have disappeared, except perhaps a few newly fledged young ones ; but Nightjars still skulk in the heaviest brackens or long shaggy heather, especially among rocks, and are seldom disturbed, as game avoid such places. Hence they are often supposed to be much rarer than they really are; but they are common enough, and on summer nights hawk about the roadsides and close round houses. Young Wheatears flick about everywhere, and Ring Ouzels cling to the cleughs and glens where they were bred. Lower down, on the burnsides, Wagtails (Pied and Grey) are conspicuous, daintily wading in the shallow water; a few Whinchats may be observed, and early in the morning the Willow-Wren still warbles a faint attempt at his spring song. The little Titlark, so well known by its irritating custom of incessantly springing right at one's feet with as much noise as a Grouse, and diving as noisily into the heather only a few yards ahead, is not generally considered a migratory bird, nevertheless it disappears from the high moors in autumn. Except in very mild seasons one seldom meets with it after the middle of October. There is another small bird I should mention, for, though I have never noticed it in August, it is sure to be present, as I have found the nest among the heather in June, viz., the Twite. It is a very inconspicuous little bird, and in the "studio sequendi" has no doubt been overlooked. Landrails remain about the rough grass on the fell edges till the middle of September or later. I have shot one as late as October 4. These birds are long in attaining the power of flight—that is, they are full-grown before their quills are fully developed. One day in August, while sitting on a grassy moor in Northumberland, I noticed a weasel emerge from a hole in a burnside and attack something which proved to be a young Landrail. I shot the weasel, and, on hunting round with a dog, found several more Landrails, evidently a brood, all nearly full-grown, but quite unable to rise, having only the blue stumps of their future quill-feathers. Swifts also in numbers seek the high ground just before their departure.

The above-named comprise most of the birds which during August have either departed southwards or are preparing to do so. The arrival of several of the autumnal migrants from the north takes place so early as (so to speak) to overlap the departure of some of the birds of sunshine. On the inland moors the autumnal migration is not nearly so conspicuous as it is on the coast. There it is usually inaugurated, often as early as the end of July, by the appearance of the Arctic Skuas and Whimbrels, both of which breed as far south as Shetland. These are followed during August by vast flights of Godwits, Knots, and other waders, many of which have come from the yet unpenetrated recesses of the Arctic regions. None of these species are at all regular visitants to the inland moors, their usual course of migration lying along the line of coast. I have, however, notes of the occurrence inland of both the first-named birds.

On the 28th August, 1878, a large flight of Whimbrels appeared, frequenting a moss-flowe, on a high moor upwards of twenty miles from the sea; and on the very same date, nine years later, several hundreds passed overhead, calling continuously. These were all arranged in V form, like wild geese. The coincidence of dates is worth recording; but otherwise the migration of Whimbrels at this season is a common enough occurrence, whether on the fells or fields, or on the sea-coast. The note of the Whimbrel is a clear, loud, tri-syllabic whistle, oft-repeated, quite audible at a mile or two's distance, and when once learnt is not easily forgotten, or mistaken for that of any other bird.

Of the Arctic Skua, the single occurrence I am aware of inland was as long ago as Sept. 12, 1854. This bird was shot by Mr. Crawhall, on Ireshope, in Weardale, passing overhead, and was in the mottled plumage of its first year. Of all the many varieties of migratory birds which in August reach our shores in such vast numbers, the above-named are the only two species I have met with on the inland moors during that month, and, to be strictly accurate, one of these was in September. The Woodcock I do not mention, as a few breed locally in Durham and Northumberland; at the same time I have only once met with one in August. The ripening crops of mountain berries attracts frequent passing visits from the migratory bands of Missel Thrushes and other birds at this season, as well as from flocks of Cushats. Packs of these latter are often feeding out on the fell edges, where Rooks and Jackdaws also revel in the abundance of caterpillars. But the resident birds are so well known as hardly to call for any remark.

BIRD-LIFE ON THE MOORS IN SEPTEMBER.

In the last chapter, I noticed the main features of moorland ornithology during August. Briefly summed up, that month may be said to witness the concluding stages of the withdrawal of most of those species which have spent the spring and summer on the hills, and the commencement of the "through transit" of others of similar kinds, which have been reared a little further north, in Scotland and the outlying islands. But of the great wave of migration from foreign lands which sets upon our coasts in August, little or no sign is visible upon the inland moors. Far different is the case upon the sea-coast. There, on the great tidal estuaries and mud-flats, the phenomenon is patent enough to the most casual observer. The dreary wastes of ooze and sand, which in June and July lay comparatively lifeless and uninteresting, by mid-August teem with countless hosts of graceful creatures—birds which but ten days or a fortnight previously, had been at home among the glacier-valleys of Spitzbergen, or Novaya Zemlya, or merrily piping amidst the desolation of the Siberian *tundra*. But of this great ornithic movement, the only indications I was able to record on the moors, were stray occurrences of Whimbrels, and of the Arctic Skua.

During September and October, we have arrivals of many of the foreign migrants, the variety of which increases as the season advances, and their consecutive appearances add a constant source of interest to shooting days in autumn. At the same time, I should add that the heather at all seasons compares unfavourably with salt-water in this respect.

Of the strictly foreign-going birds, the earliest arrival on the fells is usually the Jacksnipe. They are hardly due before October, but in the course of some twenty seasons' shooting,

I have four times met with Jacksnipes in September as follows :—

1869	September	21st	In a turnip field.
1875	,,	27th	All on the open moor.
1882	,,	23rd	
1883	,,	24th	

On the last-named occasion, there were several of them together, and evidently just arrived, for I had searched in the morning the particular spot where my dog found them at dusk in the evening on our way home. On first arrival, Jacksnipe sometimes pitch on the barest and driest places, where there is no covert but the dead and weather-bleached stalks of burnt heather. The Great, or Solitary Snipe also arrives in September, or rather it passes through this country at that date, for none spend the winter here. I have never myself met with it, and Mr. Crawhall only once in his much longer experience. This one he killed at Eshott, near Felton, Northumberland, on the 12th September, 1872; it rose before a dog standing to some partridge in rough grass, and curiously, the same shot killed one of the latter birds, unseen on the ground. The Great Snipe is a very scarce bird, far more so than it is usually considered, for many old sportsmen (and not a few young ones) imagine they have killed several. When put to the test, these supposed Solitary Snipe nearly always prove to be merely the common species, rather larger than usual. Thus their supposed occurrence as often as not takes place in winter, at which season the Solitary Snipe is never found in the British Islands. Mere weight is not a sufficient criterion, but the species may be distinguished at a glance by its underside being barred with black, these parts in the common kind being plain white. *Apropos* of the Great Snipe, an incident occurred the week I wrote this which is worth recording as showing how unreliable are most reported occurrences of scarce birds *unless obtained.* While shooting on September 15th, my setter flushed a bird which, from its size, slow flight, and general appearance, I felt certain could be nothing else than *Scolopax major*; but on following its line, presently "Nell" found it again, when I picked it up squatted in a rough tuft of grass under her nose, and found it was a *wounded Golden Plover*.

While on the subject of scarce game-birds, I may mention the Red-legged Partridge and the Quail. Of the latter, the only one I ever saw in this country was killed by Mr. Crawhall, while we were shooting together on September 22, 1870; it rose from a hill-stubble above Frosterley, in Weardale, co. Durham. Others have been obtained from time to time in the northern counties, some even in winter; but they can only be regarded as stragglers, for the Quail is a most erratic species. The Red-legged Partridge is quite unknown in the far north of England, and is not included in Mr. Hancock's "Catalogue of the Birds of Northumberland and Durham." I merely mention it here, because I once killed one in the North Riding of Yorkshire, within three or four miles of the boundary river, the Tees; it rose with some of the ordinary Grey Partridge from a field of standing beans at Hilton, near Yarm, where I was shooting with Colonel Hay, on September 21, 1877, and I have seldom been more surprised than when I picked it up and saw what it was. Red-legs, I believe, are not uncommon now in Yorkshire, but it appears they never stray across the Tees into Durham. In Portugal I have shot these partridges on high heathery ridges, not unlike many of our Northumbrian moors, and considered them excellent sporting birds, despite the bad character they bear at home. Their pedestrian powers are unquestionable; but among heather, they cannot run much more than Grouse, and when the covies are broken, the single birds lie as close as can be desired. They only require good dogs and hard work.

Having rambled so far away from the moors, and got among the stubbles and turnips, I may perhaps trespass a little further, in order to remark upon the great numbers of small birds which, on the north-east coast, are always met with among the root-crops while partridge-shooting in September. I daresay every one has noticed this feature of the season, but perhaps without particularly considering how, why, and whence these birds have come. Almost at every step they flutter up in dozens from among the turnips or "'taties," and speed away. Blackbirds, Thrushes, Larks, and

Pipits are the most numerous; and besides these, I have noticed in some seasons, very great numbers of Redstarts (all immature), sometimes also the tiny Golden-crested Wrens, right out in the middle of open turnip-fields. The quantities of birds of different kinds sometimes congregated in a single field preclude all possibility of their being the ordinary natives or residents of the neighbourhood. Unquestionably they are birds on migration, though many of them belong to species which are not popularly supposed to be migratory. There are, however, very few species which are not migratory to some extent. On examining their seasonal distribution throughout the year (which is the only true test), very few species indeed will be found to remain actually stationary during the twelvemonth; the number of such might almost be counted on the fingers.

The subject of migration has already been discussed generally when dealing with the spring season, and it is unnecessary to revert further to its general scope or character. But the phenomenon is a bi-annual one; there is the vernal movement northwards, and we have now reached, in autumn, its second and converse phase. In September the feathered world is on the move; throughout Europe, almost every individual unit is a traveller. They move in battalions or in handfuls; some traverse thousands, others hundreds of miles; there are long-winged forms that span the world, apparently without rest or effort; the Curlew-Sandpiper, for example, passes from Arctic to Tropic within a few weeks—perhaps days—and by changing hemispheres at each equinox, eliminates the element of winter entirely from its little life. Others travel more slowly, or by stages, or, on the other extreme, like the Brent Geese, cling so tenaciously to hyperborean shores, that at no period of the year are they ever fairly out of sight of ice and snow. Another section is of less extensive mobility, and a few are content with a merely local change of residence. Then, again, vertical altitude and longitudinal distance (in a north and south direction) are equivalents; that is to say the physical conditions of existence on a fell plateau of 3,000 ft. elevation above sea level in our islands, may be exactly analogous with those of low-lying lands very much further to the north, such as the South Siberian *tundras*. Thus while one bird—a Titlark, for example—which has spent the summer on the bleak uplands of Cheviot, may find a sufficient change in seeking the shelter of the littoral plains, yet the migrant instinct of another, specifically the same, but a denizen of a lower region, may not be satisfied till the owner reaches its appointed winter home in the extreme south of Europe.

There is a certain analogy between some visible traits of bird-life, and the aspirations, more or less latent, of our own kind. If it be possible to conceive a man wholly divested of all the trammels that the social and political cosmogony, and, not least, that gold have imposed upon his freedom with all the accumulated force of long custom—how would he spend the year? Perhaps, very much the same as those creatures actually do which still remain unfettered. In summer, he would probably seek northern latitudes, cruise on the verge of the Arctic ice, and retire southward before its autumnal advance, to winter in sunny lands. Such, in fact, is the custom of the fortunate few, as it is certainly the innate, and all but universal instinct in the feathered biped. Long ages of change and development have induced infinite modifications as between the different organizations of the latter; many, falling from their former high estate, no longer seek the *Ultima Thule* that once perhaps was the universal home of all; some, indeed, regard even a British summer as a thing not to be endured, and treat the Pyrenees, the Mediterranean, or the Sahara as the climatic limit of to-day—to-morrow the precession of the equinoxes may have changed all that, but the process requires twice ten thousand years.

In September the Arctic summer is over; the "midnight sun," to use the tourist phrase, has set; the ice is forming in the sounds of Spitzbergen, and will soon envelop the whole Arctic Archipelago, blocking the Kara Gates and the North Asiatic seaboard. Before the cutting hailstorms, and the approaching Polar night, the feathered world—with one solitary exception—flee to the southward. As early as July the first stray symptoms of movement become apparent at home, and in September it is in full operation—hardly a bird but by the end of that month is far to the southward of its summer position. But I must not forget that I write of the moorlands, where the winged pilgrimage is far less patent than on the tidal waters, and along the coast lines. The Whimbrels, as already stated, make straight passages and take cross-country routes ; Mr. Crawhall has shot the Greenshank and the Reeve inland, during September, and I have stray notes of the occurrence of Dotterel and Green Sandpiper; but with the exception of these, the Great Snipe, and a few ducks and other kinds to be hereafter mentioned, there is but little evidence of the passing foreigner in the hill-country.

Throughout September, the transit of Missel-Thrushes continues uninterruptedly. When driving the woods for black-game, little patches of an acre or two are often literally swarming with Common and Missel-Thrushes, Blackbirds and Ring-Ouzels. The attraction of the ripening rowan-berries, whose bright scarlet clusters are so beautiful at this season, is no doubt a retarding element, inducing the passing bands to linger a few days; but by October they are all gone. Though I mentioned the Blackbird, yet it appears to be far more sedentary as a species than most of the others referred to. It breeds in the blazing heat of the southernmost corner of Europe; yet along with the Robin, Wren, and Dipper, is one of the few small birds that can brave the utmost severities of winter on the northern moors.

The moorland districts are too high, cold, and bleak, to lend themselves kindly to corn-growing, and even in favourable seasons the little oat-crops are rarely cut till towards the end of September. At that period the stubbles, which are few and far between—often only little patches of an acre or so—are a favourite resort of the hill-game. But the capriciousness of game and some other birds is remarkable. To certain fields or particular spots they will return in spite of persistent persecution. Though shot at day after day, Partridge and Blackgame will be found regularly next morning at the same spot, so enamoured of it do they appear, and although other places, apparently exactly similar, are not frequented. No doubt, the reason is to be found in the presence of some food or other natural feature in the selected spot, unknown to us, but which is wanting elsewhere. This capriciousness is most conspicuous at the stubbles, which being few, would each, one would imagine, prove attractive to a fair quantity of game. But it is not so; perhaps one little patch is daily resorted to by a large pack, while another, to all appearances equally attractive, is invariably found blank. As the conditions change with the annual rotation of crops, the neglected field of one year may become the favoured resort of the next; so that each season the problem of their exact haunts has to be solved afresh, which adds to the pleasure of sport. The same rule also holds good to some extent on the moors above, where the burning of the old heather, and the ever varying growth of the new, each year alters the face of the hills, and consequently the haunts and habits of the Grouse.

Corn not being a natural food to Grouse, they only acquire a taste for it by its close proximity to their heathery domains. On ground where the few stubbles were far below the lowest levels of the heather, I never saw Grouse come down to the corn; their education was not advanced to that point; but in Weardale and elsewhere, where corn and heather adjoin, or even overlap, the Grouse, under certain conditions, learn to come pretty constantly to the stubbles at night.

Partridge in the hill-country, being much scarcer than in the corn-lands, are far more shifty in their habits, and bolder in their flights. They often roost a mile, or even two miles, from their basks and feeding grounds, and when put up, fly a great distance over the rough grass and bare open hill-sides; whereas, in a corn-country with its numerous hedges to hide their course, they seldom go more than a field or two. The difficulty, therefore, of finding again the fell-partridge in a wide uninclosed country is much greater, and the sport proportionately more enjoyable than among the everlasting root-crops—that is as *sport*, for, as regards shooting alone, the latter country is many-fold more prolific of results.

It is curious, when one considers the immense number of game-birds which are shot every season, how seldom they vary from the normal types. I have only myself noticed two slight instances of variation, once in a Partridge, and once a young Blackcock. The latter, shot September 12th, had each of the new black feathers in flanks and breast streaked centrally with white. The peculiarity in the Partridge consisted in its having all those portions of the scapulars and back feathers which are normally deep chestnut, quite black,

YOUNG BLACKCOCK. (1ST OF SEPTEMBER.)

as were also the bars on the flank-feathers. The bird was an old barren hen, one of three such found all together on September 4th. I shot them all, but the other two were in ordinary plumage. There is in Northumberland, a race of very dark plumaged Partridge, which is described and figured in Mr. Hancock's Catalogue, pp. 91–93.

BLACKGAME.

Of all seasons, the period from the end of August, to about the middle of October is, on the Border moorlands, the most difficult to kill Grouse. They have been so harassed and driven about by the August shooting, that they have really no fixed haunts or flights, but go flocking about, seeking safety in numbers. They sit in packs of fifty or sixty, hidden among shaggy heather, always in open unapproachable places, though not "showing" at all as yet. The weather at this season is often broken, and heavy gales and rain prevailing about the equinox, tend to increase their wildness. Later on, with fine sharp weather in October, these packs break up, and they are then, though no less wild (but being scattered about in twos and threes, and sitting bare) more possible to negotiate.

During September the young Blackgame are much more attractive than the impracticable Grouse. Indeed, their presence, especially at this season, largely compensates for the comparatively much smaller bags of Grouse attainable in those districts where both these sorts of game-birds are found together. Easy as young Blackcocks are to shoot, yet their pursuit possesses many very delightful features, both in the variety it affords after the August Grouse-shooting, and also in the changed scenes amidst which it is carried on. Whilst in August one's eye rests day after day upon an almost unvarying, unbroken sea of purple heather, glorious in its fullest bloom, with its golden pollen streaming away in a little cloud to leeward of the course of dog and man ; now our sport lies amidst widely different scenes, no less wild and hardly less beautiful. Stretches of rolling prairie-land, of rough grass, rush, and bracken, interspersed here and there with straggling patches of natural birch and hazel, take the place of the heather ;

and instead of wide-spreading moors, one now rambles along tortuous little cleughs, shaggy with lichen-covered birch and rowan tree, or up the rugged course of a steep-sided rocky glen—the favourite haunt of young "grey," and many of which are among the most exquisitely wild and charming nooks ever carved out by Nature. In these sequestered spots, as a September sun shines brightly through the scattered birches, upon the masses of bracken and variegated foliage below, amongst which the setters are bustling about, their russet coats in sharp contrast with the dark rushes and paler fern, surely one has as fair a scene as eye need wish to rest on.

Young Blackgame are among the slowest of game-birds in attaining maturity. They are hatched early in June, but cannot be considered full-grown till the end of September, and during their four months of adolescence are certainly the "softest" and most tender of all the game-birds—a curious contrast with their strong and hardy nature when adult. Even when half-grown it is quite common to see a young Blackcock, if put up two or three times on a wet day, become so draggled and exhausted as to be unable to rise again. The habits of young Blackgame are precisely analogous with their tardy bodily development. All through their protracted adolescence, and during August and September, they are the very tamest of birds. Then all at once they appear to gain a sudden accession of strength and wildness; their timid skulking nature is discarded along with the weak, little, pointed, ruddy tails of their nestling plumage, and in a few weeks, even days, the young Blackcock, from being the tamest, becomes the wildest of all our game-birds.

To shoot the young Blackgame in August, when they are hardly bigger than Quail, and before the cocks can be distinguished from the hens, is not only so unsportsmanlike a proceeding, but it is so suicidal a policy, that one might expect few people would be guilty of it. Yet I regret to say it is far too prevalent a custom in many of the Border districts. Hundreds of the wretched little fledgelings are massacred on the "20th," many even before that date, apparently from sheer greed on the part of their murderers, or from a desire to deceive the ignorant by boasting of the numbers of their bag. Thus the excuse is sometimes made, with a vacuous grin, "You know, somebody else might get them." Well, let *somebody else* have them, and welcome, if he cares for them, and can derive either pride or pleasure from turning out a sackful of bloodstained pulpy remains, useless alike for sport or the spit, and far more resembling dish-clouts than game-birds. The worst offenders in this respect are the semi-respectable gentry who rent "moors" of, say, 100 (!) acres, with the deliberate intention, as soon as the adjoining owner is out of sight and hearing, of slipping over the boundary, where they can pillage and massacre to their heart's content. From such one expects nothing better, and one knows how to deal with them; but truth compels me to add that the malpractice is not confined to them.

By the middle of September the young Blackcocks are nearly full-grown, and about three parts black, with spreading tails. At that period they separate themselves from the young Greyhens of the brood, and for a time become quite solitary. Being then scattered singly over a wide extent of rough country, they are less easy to find than to get at, for, though nearly full-sized, they lie extremely close in beds of bracken and rushes, or in the "white grass" or patches of heather. Towards dusk they begin to feed on the seeds of rushes, especially the "spratt" or flowering rush, and being then temporarily gathered together, are much wilder than during the day. They continue "on feed" till it is quite dark.

This (mid-September) is the season when young Blackgame afford by far the finest sport over dogs; for though they lie close and offer easy shots, they require a great deal of hunting for, and a bag of perhaps eight or ten brace of well-grown handsome young birds, varied by a few brace of moor-partridge, and an odd grouse or two picked up on the outskirts of the heather, is a very satisfactory day's work. This is the time to gather up one's "crop" of Blackgame.

Their next stage is to assemble in packs about the end of September or early in October, sooner or later according to the season, cocks and hens together, often a hundred strong. These packs make descents at daybreak and dusk upon the scattered patches of oat-stubbles in the valleys and spend the day, perhaps miles away, among the hills, preferring the outskirts of a straggling birch or pine wood with plenty of rough ferny bottom, or in some rugged cleugh where a little tumbling hill-burn has cut itself a deep tortuous ravine, whose steep rocky sides are overgrown with rank heather, bog-myrtle, and bracken, and studded with stunted birch, alder, and mountain-ash. They now become watchful and wild, and are difficult to handle comprehensively; the young cocks being almost indistinguishable from old ones, and having finer "tails." In point of numbers, one-half the quantity one could get a fortnight or so ago should now be considered a fair bag.

The old Blackcocks at this season are still in the moult. At the beginning of the season (August 20) few of the old

YOUNG BLACKCOCK. (SHOT END OF SEPTEMRER.)

cocks have the slightest vestige of a tail; *i.e.*, the short blood-feathers of the nascent tail are entirely hidden by the upper and under coverts which meet beyond them. At that period they usually lie very close, skulking in beds of the heaviest bracken, as though ashamed of their ragged condition. Now and then one shoots an old Blackcock in August, which, probably from backward condition or other cause, still retains some, or even all, of the long curved tail-feathers of the previous year, of course very worn and ragged. Such birds are usually very large—three or four-year-old birds—but they scale less than smaller tail-less old cocks.

During what I may call the "stubble period," *i.e.*, from about the middle of October till towards the end of November (varying of course according to the date of the harvest),

Blackgame are rather scarce on the higher moors. In September they find abundant food on the hills in Nature's crop of seeds, and the mountain berries, blaeberries, cranberries, &c.; but as these become exhausted, and the corn-crops being by then usually led out of the fields, large numbers of Blackgame leave the higher hills and resort to the corn-lands in the lower ground. Later on, as the stubbles are ploughed up, they return to their hill-haunts, usually about the end of November, and their food then consists largely of heather, and the small plants which grow in (or rather form) "old grass land." The following are the contents of the crops of four Blackcocks shot November 3. The first, killed out on the fell, contained heather alone. A brace shot on the fell-edge, half heather and half aromatic grass-plants, the former uppermost; these birds had probably been disturbed in the hay-fields below, and had come out on to the heather to finish their dinner. The fourth, shot on grass-land at dusk, had fed entirely on grass-plants—trefoils, sorrels, sedges, &c. Several others examined contained much the same food; sometimes a few dozen oats at the bottom of the crop, though none might happen to be grown within miles of where the bird was killed.

All game-birds feed very low, crouching along on the barest ground. It is surprising how easy it is sometimes to overlook even so large and conspicuous a bird as a Blackcock when feeding. A pack of them may be feeding on nearly bare grass, slowly advancing with all heads and tails down, and yet may be overlooked, or perhaps taken at a careless glance to be only a lot of mole-hills.

There is a remarkable feature in the habits of Blackgame in mid-autumn, the cause of which I have never been able to make out. I can find no explanation of it recorded, and indeed it seems quite inexplicable. I refer to the distinct display of amatory instincts which occurs in October, and in mild seasons even later. On wet, foggy mornings in particular, one hears the old Blackcocks "crooning," "bubbling," and "sneezing," as excitedly as on a fine day in early spring. With a glass, I have watched one surrounded by his harem, strutting round some bare little knowe in the fullest "play," his neck swollen, tail expanded erect over his back, and wings trailing—truly a most remarkable spectacle in October. Whether it is merely a chronological miscalculation, or arises from some specific cause, the origin of which may be lost in the mists of the remote past, the instinct certainly exists, and, for want of a better name, I will coin the word "pseudo-erotism," to designate it. Nor is it confined to Blackgame. Grouse conspicuously, and Golden Plover to a certain extent, are affected by "pseudo-erotic" instincts; and we all know how busily Rooks destroy and repair their nests in November. I have also noticed small Gulls in that month, apparently Black-headed Gulls, revisiting their moorland breeding haunts. So strong is this instinct in Blackgame, that in December, when the snow lay a foot deep, I have observed the young cocks (which had then apparently arrived at maturity) similarly dancing around and among a pack of grey, the latter as usual apparently utterly heedless of the performance, though I should add it was not nearly so "impressive" as that of the old cocks in October and November.

During the concluding months of the year the habits of Blackgame do not materially alter, except as they are influenced by the weather. They are now firmly established on the high moors, and we have in Northumberland a considerable immigration at this period, probably from the higher bleak fells about the Scottish border, and from their great strongholds in the wild hills of Roxburghshire, Selkirk, &c. It is no unusual occurrence to see more Blackgame on Dec. 10 than could have been seen at any previous period of the season. They now select certain fixed haunts, usually some high flat-topped ridge, where patches of short sweet grass are interspersed among the heather, to which they constantly resort, and where perhaps thirty or forty Blackcock and a great number of grey may always be found. Heavy gales of wind and rain-storms often drive the Blackgame off the hills at this season to the shelter of the wooded valleys and cleughs below—but not always; for in attempting to describe the habits of birds, it is difficult to lay down absolute rules. So many and such varied causes (some perhaps unknown, or impalpable to us) affect their habits and movements, that it is unsafe to write dogmatically, and nearly all observations should be made and read in a general sense. Thus, we sometimes obtain excellent sport by driving the woods and gills the morning after a storm. But on other days, under what appear precisely similar circumstances, hardly a bird has been found in the shelter.

Not always, however, do the birds and the hill-farmers enjoy the luxury of a mild winter. Often the "genuine article" is ushered in in November, and storm follows storm till the brown heather disappears entirely beneath the universal covering of dazzling snow. Under these conditions, the Blackgame (though many of them still cling to their chosen hillocks above with surprising tenacity) are generally to be found congregated in the wooded valleys below, where dozens of them may be seen perched like Rooks on the bare birches and hawthorns. Here they feed chiefly on haws, and on the budding shoots or "tops" of birch and alder. They also get a certain amount of heather in places where the wind has drifted the snow from the weather-slopes of the hills, or where sheep have been feeding. Blackgame stand prolonged snow-storms with great hardiness, and show little or no falling off in condition; but in the snow-storms of December 1882, a friend tells me he found many Grey-hens dead and dying on his farm, even in the stackyards. The crops of nearly all these birds were quite full, and death was ascribed to their inability to obtain the necessary supplies of sand, gravel, &c., required to promote digestion.

A question has been raised as to whether Grey-hens breed in their first spring, and I have even heard it stated that they do not breed till their third year. This, of course, is a matter very difficult, or even incapable, of direct proof; but so precarious are the lives of game-birds that, if handicapped so severely, I imagine they could hardly manage long to maintain the "struggle for existence." My own observation leads me unhesitatingly to say that Grey-hens *do* breed in their first year, or at least as many of them as have the opportunity. The number of Grey-hens without broods never appears greater than can be accounted for by the fact of the species being polygamous.

BIRD-LIFE ON THE MOORS IN OCTOBER.

THE month of October is often inaugurated by the appearance of the Wild Geese, which usually arrive soon after harvest. As early as Oct. 1, I have seen the large Grey Geese passing high overhead in noisy skeins and V-shaped lines. Their course of flight is almost invariably to the westward; they seldom alight, and, on the rare occasions when they do so, are so extremely cautious in their choice of a resting-place, and so incessantly vigilant, that they rarely allow an "advantage" to the gunner. Hence they are seldom killed. I have never, myself, succeeded in bagging a Goose on the Northumbrian moors (though from no lack of effort), and cannot, therefore, speak confidently as to their species. Doubtless, however, these will be similar to what are usually obtained on the coast, viz., the Pink-footed and Bean Geese most common; the smaller White-fronted kind less so, and the big Greylag the rarest, if, indeed, it ever visits us at all. On Oct. 13, 1878, a small flight of thirteen Grey Geese took up their quarters on a moorland lough, which was partly frozen, and remained some days, sitting on the edge of the ice, where their watchful sentries defied all attempts to outmanœuvre them by day and night.

The moorland lakes, or "loughs" of Northumberland (or *Scotice*, "lochs"; but, being south of the Border, I prefer the local name, which is pronounced "loff,") are, as a rule, most unfavourable places for approaching wildfowl. Many are situated high up, among the crests of the heathery hills, where there is seldom a vestige of cover on their banks, not even a fringe of rushes, or any bush or shrub, higher than heather or bog-myrtle, to cover a stalk. Some are simply open peat-holes in the middle of a dead flat bog, their surface not a foot below the general level of the surrounding "moss," and, on these, direct approach on wildfowl is utterly out of the question. Others, lying in shallow basins among the hills, with the bare heather sloping right down to the water's edge, afford hardly any greater advantage; though, in these latter, there is usually a "syke," or broken gully, at the over-flow, sufficiently deep to enable one to creep within reach of the water at that point. These loughs usually have a firm bottom, either peat or gravel, the latter here and there interspersed with beautiful patches of silver sand; but there is no shore where fowl can sit dry, the water being deep right up to the steep banks.

Wherever a section of the peat-formation is exposed, trunks and remains of ancient oak, and other trees, are visible, even up to 1,200ft. and upwards, the relics of a long past age, when these now open, treeless moors were covered with forest. These remains appear to be analogous to the upright stems known as "sigillaria" in the coal seams of the older carboniferous period.

Years ago, before the celebrated wildfowl resort at Prestwick Carr, in Northumberland, was reclaimed, the geese from the carr used regularly to resort to this and other moorland loughs for a wide radius around to roost, and many of the hill-farmers have anecdotes of incidents which occurred when they went to lie in wait for the arrival of the geese at nights. But now all that is changed. Corn grows where only waste marsh and water formerly existed, and except on a few such occasions as the above, I have never seen the geese condescend to remain a single hour on the moors. I need hardly say that the Brent Goose never appears inland. It is, essentially, a salt-water fowl, and, on the coast, more numerous than all the other kinds put together.

By the middle of October the regular winter birds begin to put in an appearance. Woodcocks, Grey Crows, Redwings, and Fieldfares arrive *en masse* about the 20th of the month. The Woodcock, on arrival, pitch down among the heather, often far out on the open moors, though never in any numbers. The earliest of these distinctly foreign birds which I recollect finding thus, was on Oct. 2.

In 1880 I witnessed an extraordinary immigration of Fieldfares on Oct. 23. While lying waiting for a "drive" of some ducks, at the edge of a lough, suddenly several thousands of Fieldfares appeared, flying south-west, and quite low over the heather, many passing within a foot of my head as I lay concealed. They uttered, continually, a peculiar low single pipe, quite different to any note I ever heard Fieldfares make before or since. For some days after this, the fells were "grey" with them, sitting about on bare (burnt) places, especially on stones. My brother had a similar experience with Redwings on Oct. 12, 1882. It was a densely thick, foggy day, the mist driving across the hills, and a little wood, which they had just tried out for Blackgame, suddenly became filled with Redwings, which continued to arrive through the fog in hundreds, keeping up a constant chirping chatter.

Neither of these birds spend the winter on the hills. There are neither trees nor hedges on the bleak moors to their taste, so they quickly pass on to more cultivated regions. Redwings, especially, make a very short stay; they usually arrive a week or ten days before the Fieldfares, but only spend a day or two to rest, feeding on the berries, and in moist meadows. In mild seasons a few of the Fieldfares occasionally remain throughout the winter.

Simultaneously with the arrival of these winter birds, occurs the departure from the fells of the last of the less hardy ones. By the middle of October the Skylarks, which a fortnight earlier swarmed about the stubbles (many of these, no doubt, on passage) have disappeared, and the Titlarks follow them at the end of the month. I have a note made some years ago, that in a week's shooting on the "white grass," at the end of October, we only observed one of the latter, where in September hundreds would have been seen. The Ring-Ouzel also departs with the fall of the leaf, though I have noticed a single straggler lingering as late as Nov. 13; he flew out at the end of a wooded glen, where we were driving Blackgame, right in my face, his white gorget plainly distinguishable.

In 1880, on Oct. 4, I noticed a duck on Darden Lough,

GOLDEN PLOVERS. (WINTER.)

To face page 96.

Northumberland, which completely puzzled me—a big, black diving duck, along with a dozen Mallards. As the stranger continued diving, remaining under water for half a minute at a time, I easily managed to "run down" on him, and pinioned him by a long shot as he rose; but after that, it cost me an hour's hard work, and nearly a dozen cartridges, ere the red paddles turned upwards, so hard was the bird, and so quick and determined a diver. This duck proved to be a Velvet Scoter, an adult female, having a grey-speckled breast, and two curious patches of white on either side of the head, one at the base of the upper mandible, and a larger and more defined patch on the ear. It weighed 3lb. 2oz., and the crop contained only gravel. This was a strange bird to find at a hill-lough, far inland; for the Velvet Scoter is essentially a sea-duck, and its occurrence here is mentioned in the 4th edition of Yarrell's "British Birds."

Hardly less remarkable is the occurrence, at this same lough, of the Sheld-Duck, another marine species. On Nov. 20, 1877, while we were grouse-shooting, my brother and Mr. Browell reconnoitred the lough, when seven Sheld-Ducks rose at the far end, and, deliberately flying right over the guns, paid the penalty of their innocence by losing half their company, three being killed and a fourth wounded. They were all immature, and it was difficult to persuade the worthy villagers of Elsdon that these gaudy ducks were not escaped stragglers from some private pond or ornamental water. Another sea-duck of which I have a single instance to record on the inland waters is the Scaup—a young drake shot in November, 1875. Of course the occurrence on the inland moors of all these purely sea-ducks can only be regarded as exceptional, and the record extends back over a considerable number of years.

The Goosander is another not infrequent autumn visitant, but it rarely appears on the still waters of the moorland loughs (which contain no fish), its preference being for running waters and the larger streams, such as Redewater, Coquet, and, more specially, Tweed. This handsome duck differs from the closely allied Merganser in being essentially a fresh-water bird, feeding on trout, and only exceptionally appearing on the coast in hard weather; whereas I have never seen the Merganser in autumn away from the salt-water. Only once have I observed a Goosander upon the hill-loughs, and that one escaped, in all probability purely by virtue of the admirably conceived protective coloration of its plumage. For, large and conspicuous as a Goosander drake appears, yet his black and white plumage assimilated so perfectly with the rippling water (it was a bright colourless November day) that, though I had carefully "glassed" the lough before showing in sight, I failed to detect anything on its surface. But on my appearing he at once rose from the water, and after circling round several times (but always out of shot) he took himself right away.

But of all the foreign-going ducks, by far the most regular winter migrant to the Border moors is the Golden-eye, whose usual date of arrival is the concluding days of October or the early ones of November. Next to the Mallard, it is the commonest of the duck-tribe on the inland moors. It appears very regularly within a day or two of the 1st of November, singly or in twos and threes, the largest number I have seen together being seven. Golden-eyes are easily distinguished from Teal at any distance by their white wing-spot, or "speculum," and by their incessant diving. I mention this because few keepers discriminate between the two species, yet different tactics are advisable to secure them. When inland, Golden-eyes are the simplest of all the duck-tribe; so much so, that on seeing some on a lough I always feel sure of getting a pair or more. This is the more remarkable, since on the coast they are among the wildest of wildfowl, and I have always found that, to attempt to punt to these ducks, or to Mergansers, by daylight (which, by the way, is the only chance (?) they offer, as all these day-feeding diving-ducks are safe enough out at sea by night) is just so much labour lost. On the moors, however, just the reverse is the case. It is only necessary to creep within shot of the water, and, on sending a man round, the Golden-eyes will deliberately fly or even swim up to the concealed gun; whereas Mallards, under similar circumstances, would at once, on the appearance of a human being, rise perhaps a hundred yards in the air, and probably not stop again within several miles. Even after being shot at, the Golden-eyes will often continue to circle round, and sometimes return to pitch among their defunct companions; so that it is worth while, after a shot, to remain *caché* for ten minutes or a quarter of an hour.

All the Golden-eyes I have shot or seen shot on the moors have been in what is considered the female or immature plumage,* though many are unquestionably drakes, as is shown by their weights thus:

Young females average . . .	1lb. 4oz. or 1lb. 5oz.
Adult females average . . .	1lb. 12oz.
Young drakes average . . .	2lb. 2oz. to 2lb. 4oz.

The irides of both the latter are golden, those of the smaller birds (the "Morillons" of Colquhoun) being brown. The adult duck is much lighter coloured on her wings, the coverts and scapulars being spotted irregularly with white, and her neck is also much whiter. I have always found them excellent eating, the flavour resembling Wigeon, but they are less oily; however, I am no judge on epicurean matters.

In 1884 I found five Golden-eyes on October 16—the earliest arrival of these winter ducks I ever noted. The keeper drove them to me, four ducks and a drake. I killed a duck and the drake, right and left, as they came overhead, stone dead, and a highly satisfactory "souse" they came down with! Weights, duck 1lb. 4½oz., drake 2lb. 1½oz.

The peat loughs do not appear congenial to the tastes of the Tufted Duck, and it is rarely found thereon—much more so than one would expect from the circumstance of its breeding in the district. I have observed these ducks late in spring on several occasions, but not in autumn: the Pochard I have never come across inland then, and the Wigeon but seldom.

During October, Teal are more numerous than in any other month of the year—no doubt on passage (I have observed

* I shot an adult drake Golden-eye on the Tweed, Oct. 19, 1888; at which date, the full plumage had not been completely acquired—the white cheek-patch and neck being still slightly obscured with dusky feathers—a phase of plumage rarely met with here.

their arrival in Portugal on October 22). They frequent small rushy pools out on the moor, having special predilections for certain spots, to which they resort year after year, though the individual birds are annually killed there. They usually sit close when one has the luck to come across them (pairs or single birds), unfortunately not very often. By the middle of the month the young wild ducks begin to "show" on the open water of the loughs, instead of skulking among rushes, or under the long overhanging heather on the lough sides, as they do in August. They also begin to feed further afield, and come down to the patches of oat-stubble, where they sometimes feed with the Blackgame. While waiting for the latter I have seen the ducks circling round to reconnoitre, even before dark. By the end of the month we often have a pack of a hundred, to a hundred and fifty or more, on Darden Lough, where they remain by day pretty constantly throughout the season, and, truth compels me to admit, in but imperceptibly diminished numbers. These are not, in my opinion, foreign birds, but an aggregation of many local broods, perhaps all that have been bred in the wild country for miles around. In all temperate countries there appear to be two well-defined and distinct races of *Anas boschas*—(1) the native-breeding birds, which are non-migratory, remaining throughout the year; and (2) the foreign contingent, which migrate from northern latitudes to spend the winter only. The two races are distinguishable by their different types. The foreigners are of a lighter and more slim build, while the sedentary race, by long desuetude, have lost the power of far-sustained flight, having gradually become heavier in body and no longer adapted to perform lengthened migrations. Thus, while our heavy moor-bred Mallard drakes scale over 3lb., the foreigner, with exactly equal width and expanse of wing, only reaches some 2lb., or at the most 2½lb. These latter, however, are seldom or never met with on the inland moors, their predilections being for the coast and tidal estuaries, where they are the staple fowl. The heavy ducks, on the other hand, are rarely found on salt water, except in winter, when "frozen out" of their moorland haunts by severe weather.

GOLDEN-EYES—OLD DUCK, YOUNG DUCK AND DRAKE. (NOVEMBER.)

To face page 100

During the latter half of October, a marked change becomes observable in the habits of the strong and wild descriptions of hill-game. The Grouse, which during September have been congregated high out on the hills in big, shifty, inaccessible packs, cowering, vigilant but invisible, among the heather, now disperse into small congeries of a couple to half a dozen birds, which sit boldly conspicuous on open "white ground" and bare knowes. At the end of October one no longer expects to get point shots at young Blackcocks, and, as already mentioned, the Mallards (the drakes having acquired their chestnut breasts, and rapidly assuming their glossy green heads) appear freely on open waters, instead of skulking in shelter. In short, all the strong wild birds, which are now attaining their full feather and beauty, begin to show more boldly. They no longer seek a delusive security in concealment. Such habits were natural enough with half-grown, half-plumaged poults early in the season, or ragged, moulting old birds; but now, with increasing strength, their former devices are thrown aside, and they sit bare and conspicuous on hillock, knowe, or lough, confident in their keen instinct of self-preservation and in their powers of wing and eye to keep themselves out of harm's way. Different tactics must now be adopted to secure them from those which in August were wont to fill the bag. Dogs are no longer any use except to find dead, for in Northumberland or Durham Grouse could as easily be secured by a process of *habeas corpus* as over dogs at this season.

Grouse towards the end of October become very noisy, especially just after their "morning flight," which takes place at, and even before, the break of day. At that hour, just as the first streak of dawn appears over the eastern hills, they commence their matutinal movements; and on a bright frosty October morning it is delightful to hear the "concert" they keep up. From every hillside and heathery knowe ring out sharply their clear loud notes, often before it is light enough to see, and with a variety of intonation which is surprising, and, I may add, most exhilarating to a sportsman's ear.

With fine weather, Grouse are mostly in pairs by the second half of October, or in small lots of four, six, or eight, which are also composed of pairs. That this is the case is most easily seen when stalking (or what is called "edging") Grouse, or, better still, when "carting" to them. Then, the courtship of the Grouse cock, and the coquettishness of his mate are conspicuously visible—even the *amantium iræ* are observable, and very amusing. Every now and then the hen Grouse dashes away, followed at once by her lover, and the chase continues for minutes at a time. Round hillocks, along sinuous hollows, now low on the heather, then high in the air, the pursuit is carried on with intense energy—the hen often dodging downwards or sideways as though a falcon were in pursuit—the while the low soft spring note is constantly repeated. Than the old Grouse cock at this season there is no more beautiful object in Nature, as he proudly stalks over the short heather, with head and tail carried almost equally erect (for he is most particular not to let the latter get wet with the melted "rime"), his steely-sheened plumage, bright scarlet comb, and chestnut throat; only a yard or two beyond is his speckled mate, but crouching low among the heather, she is almost invisible to an unpractised eye.

Few things have struck me as more remarkable than, first, the frequent inability of a novice on the fells to distinguish a Grouse which is close at hand and apparently in full view; and, secondly, the extraordinary acuteness of eyesight which is developed by constant practice in those who live among the hills. "Well, Sir, I thought I just kenned the turn of his neb," or "the red on his kame!" is the reply when one asks how in the name of all that is wonderful one's companion has detected a Grouse far away and low among the heather. And one mentally measures the distance from the original point of view with something of vexation in one's feelings at what appears the utter hopelessness of ever attaining such keenness in the "visual ray." Closely, however, as the plumage of the moor-game approximates to the brown colour and broken "quality" of the autumnal heather, yet there is a sheen on it, and a wholeness of colour, which is distinguishable to a practised eye which knows what to look for.

To resume: I have already referred to the above-mentioned strange recrudescence of amatory symptoms in mid-autumn in a former chapter, when writing of Blackgame, and therein christened the phenomenon pseudo-erotism. So far as I recollect, neither St. John nor Colquhoun mention this apparently unaccountable trait in the character of certain birds at this season, beyond a passing reference, in the former author's well-known "Sport in Moray" (p. 221), to the habit of Blackgame to assemble, and the cocks to "call," during October. While in the Highlands, I have made inquiries from gamekeepers and others on this point, but without being able to ascertain that the phenomenon was at all known to them. Possibly this may arise from want of observation, for that first-rate sportsman Lloyd, in his "Field Sports of the North of Europe," describes the habit as observed in the Capercaillie in Norway—(and, if I remember aright, in the Blackcock as well). Here is a further note on the subject relating to the Golden Plover: "Oct. 31, 1882. To-day, in fine warm sunshine, observed the Golden Plover persistently chasing each other, and repeatedly uttering their pretty love-note of the spring. There was a large flight of them, perhaps two hundred, and evidently in exuberant spirits; now high up in the clouds, then suddenly darting down in a hundred curving lines, like falling stars, right to the very heather, whence they rose again, reuniting into close order in the sky, when the pack would again shiver into atoms, dashing headlong downwards in every direction."

In reference to the spring-note of the Golden Plover, it is often surprising at this season to hear the absolutely perfect imitation of it which is produced by the common Starling. In some old trees before my front door a colony of these birds have their head-quarters, and they frequently amuse me early on a bright October morning, while lighting the matutinal pipe and preparing for a start, by their exquisite reproduction of this soft gurgling note, and also of the loud weird spring whistle of the Curlew. The latter, at any rate, they cannot possibly have heard for many months, and the Plover's note always appeared to me absolutely incapable of imitation. The Starling's memory must be as good as are his powers of mimicry.

The Grey-backed Crows are at this season most inimical to sport. They hunt the heather as regularly as a setter, and invariably put up every Grouse they find, checking their flight whenever they come over a game-bird, apparently to see if it happens to be wounded. In this way they are sometimes the means of one's recovering a wounded Grouse, but far more often they have cleared a whole hillside of moorgame which one had laboriously driven in, in hopes of filling the bag. Why unwounded Grouse or Blackgame should fly from these Crows is not apparent—for the latter are quite incapable of injuring them—but they invariably do so ; and even Rooks make feints at Grouse, which always put them up.

The latter birds (Rooks) are extremely fond of a feast upon Grouse when procurable, and daily search the sides of the old coach-road which crosses the Border moors on its way from Newcastle to Edinburgh, and along which a telegraphic line is stretched. This line at present consists of nineteen wires—a perfect trap for birds, and the damage it causes to bird-life is incredible. I have heard it estimated by farmers and shepherds (and believe they are not far wrong) that more Grouse meet their deaths annually from these mischievous wires than are killed by all the shooters on the moors around. The nineteen wires cover so much space, and being stretched at exactly the usual height of the flight of game-birds (and especially of their morning flight, when in the indistinct light the wires are wholly invisible) that they cannot fail in their destructive work, and occasionally a pack is cut down by wholesale. It should be remembered, too, that this destruction is going on at all seasons of the year. It is no exaggeration to say that the roadside is at certain seasons strewn with remains. Besides Grouse, I have picked up Blackgame, Partridge, Curlew, Golden Plover, Snipe, Peewits, and other birds. Every morning at break of day come out the marauding bands of Rooks from the lowland woods, reconnoitring along the roadside, and feasting on the dead and dying. I meet them regularly at dawn as I walk across the moors to catch the early morning train.

In order to give some idea of the mischievous nature of these wires, and of the cruelty and ceaseless suffering they occasion to the moor birds, I copy the following few extracts from my shooting diary: "Oct. 6. Found to-day four Grouse which had been severely damaged by flying against the telegraph wires on Elsdon Hillhead. Two were already dead, and pulled to bits by the Crows. The third had evidently received his wound late the night before, and the blow had completely carried away his crop, which at that time would

PEEWITS—A MIDDAY SIESTA.

be full of heather. The poor bird had been hungry this morning, and, regardless or oblivious of having no crop, had been feeding, his throat down to the huge gash being crammed with heather shoots. I never saw anything more pitiable in my life. This bird could still fly, but very weakly, and could not possibly long have survived. The fourth Grouse had been injured some time before. He also had received a horrible gash across the breast, but it appeared to be slowly healing. His breast was bare of feathers, and the old skin was hard and yellow, a mass of clotted blood remaining in the cut. This bird flew nearly half a mile when put up by the keeper (driving), but was very weak and unwilling to rise." "Oct. 17. Every day this week, when shooting near the telegraph lines, we have found Grouse either killed or severely injured by the wires, and to-day I shot a Grouse in a horribly mangled state at Laing's Hill, several miles away from the line." The above are sufficient illustrations of what I have stated, though it would be easy to adduce hundreds of similar instances. Surely, in these days of ultra-humanitarianism, of R.S.P.C.A. associations, and of "Wild Bird Protection Acts"—when a maudlin sentimentality comforts itself by fining a poor man for shooting a wild goose in March, or for overworking his horse, on which perhaps depends his daily bread—surely, in these days, the wanton cruelty and useless waste above described (carried on for a national profit) should not be permitted. But then these cruelties are *not seen;* they only occur on the remote hills, where no one witnesses them save shepherds.

On October 8, 1879, I came across what is now a very rare bird on the Northumberland moors, a Hen-harrier, an adult male, pale blue. He had struck down a Grouse, obviously on the wing, and was busy eating it in the bottom of a deep black ravine or chasm in the peat, into which the Grouse had fallen, when my setter pointed the pair from the opposite side. The Harrier rose from almost under my feet with a loud scream, his yellow claws dangling below him; but in my extreme anxiety not to blow him to fragments (I was shooting with a No. 10 gun, full choke), I let him well away, when "skirling" about in half a gale of wind, I clean missed him. The head of the Grouse had been completely severed from its body, and lay some 10 ft. or 12 ft. lower down the ravine. It is worth mentioning, that in beating up-wind to where we found the Harrier (which was on one of the highest ridges of the fell), the Grouse had been "lying" well—a most unusual occurrence at that season—and my brother and I had just killed four or five brace over dogs. For this I think it is probable we had to thank the Harrier, as, from the position of the dead Grouse and its head and other circumstances, he appeared to have hunted the ground up-wind, just in advance of us.

On one other occasion only have I personally observed the Hen-harrier among the solitudes of the Borders, this was on

June 1, 1884. I raised this hawk, which was again an adult male, on the outskirts of a flat low-lying bog, and afterwards followed and put him up several times, each time carrying some prey in his claws. From his demeanour and other circumstances, I felt certain he was one of a pair then nesting in the bog, which was of great extent, a perfect sea of level rushes, and the beau ideal of the favourite resort of a Harrier.

I have never succeeded in actually finding the Harrier nesting on the moors, but have frequently had accounts given me of its doing so, which I have no doubt are correct. In view of the fact that the Hen-harrier is regularly migratory as a species, being strictly a winter visitant to Southern Europe, and having a breeding range extending beyond the Arctic circle, it is hardly surprising if a few pairs do occasionally find their way to nest among the wide expanse of fell and flowe extending all along the Borderland, and that despite all the persecution of gamekeepers. There are a few favoured localities which I could name, but purposely refrain from doing so from dread of the "collector." It is hardly too severe to describe certain of this class as perfect pests in their wholesale depredations. Under the cloak of science, some even carry on what is nothing better than a trade in birds and their eggs. The eggs of such kinds as breed *gregatim* are swept up by wholesale, while the few remaining survivors of our rarer birds are threatened with extermination through the bribes which are offered to shepherds and keepers, often in direct opposition to the wishes of their masters.

The Peregrine is now a very rare bird in Northumberland, and it is only at long intervals that one has the pleasure of observing its dashing flight. The wild moory hills and rugged crags which Nature assigned for its dominion, and where in years gone by this fine Falcon, together with the Buzzard, the Raven, and the Harrier, regularly nested, will in all probability soon know it no more. Rightly or wrongly, man has usurped the functions of Nature in adjusting the balance of life (I do not mean to imply here any opinion on the subject, and much can be said on either side), but it does appear regrettable that the fate of the few survivors of these noble aborigines should in some cases be left at the mercy of ignorance and prejudice. The present status of the Peregrine as a resident bird in Northumberland, and along the Borders is most precarious: it can still be so described, but, owing to the ceaseless persecution in spring, a whole season will sometimes elapse without so much as seeing a Peregrine on the hills.

The Buzzard, too, has disappeared. I have never once seen one on the wing, and not a single pair now nest in Durham or Northumberland. The few Buzzards that do occur, are generally met with during the month of October, but these are usually of the northern type, or Rough-legged Buzzard (*Buteo lagopus*), of which species I have examined three or four shot at this season. These are no doubt merely belated stragglers from the ranks of the great migrating bands which, on the approach of winter, pass southwards from the shores of the Baltic and from Northern Europe, but whose course lies in the main to the eastward of our islands. I have also seen one specimen of the Common Buzzard shot at this season, and a Goshawk (immature) killed in November.

Merlins are tolerably abundant on the Border moors, where a few pairs nest in spring among the heather; but it is in September and October that these little falcons are most numerous—chiefly young birds, which prey on Larks and other small birds along the fell edges. These are no doubt on migration, but we have some throughout the winter, for I have records of them in November, December, and January. Adult males are comparatively scarce; their blue backs show in conspicuous contrast with the dark heather when seen flying over it, but harmonize admirably in colour with the big grey boulders on which they are so fond of perching.

At the end of the month the Grey Wagtail (*Motacilla boarula*) is noticeably abundant; their numbers decrease as the winter advances, but in mild seasons a few remain about the burn-sides during December and January. Peewits also disappear from the high moors during winter; at this season (end of October) they are often congregated during wet weather on the bare black ground where the heather has been burnt, and where they feed all night. Frost or snow at once drives them off the hills.

The oceanic birds, such as Solan Geese, Petrels, and Little Auks, seem peculiarly liable to get driven inland after stormy weather at sea. One often hears of instances of their occurrence in most unlikely localities. Thus a young Solan Goose was caught alive near Elsdon on October 31, 1883; it was in an exhausted condition, and did not long survive; but that was hardly surprising, since the only food he was offered was cold mutton!

The following cutting from a local paper evidently refers to another occurrence of this species (a young Solan Goose) inland, and is sufficiently amusing to deserve insertion:—

> A wonderful bird has been found by a shepherd near Kirton-in-Lindsay, Lincolnshire. It is about the size of a Turkey, dark grey speckled, web-footed, quite amphibious. The naturalists of the district are quite puzzled, some thinking it to be a Northern diver, and others a Vulture escaped from a ship, or driven away by the gale.—*Newcastle Daily Journal, October* 22, 1886.

Small wonder, surely, that the naturalists of the district should be puzzled with a *Vulture* which was web-footed, and "quite amphibious"!

CARTING TO MOORGAME.

THE month of September, as already mentioned, is one of the worst periods of the season for killing grouse: after the August bombardment they are harassed, restless, and packed, and moreover they do not yet "show" at all. They are in fact wilder, or at least less negotiable than at any other season. The Blackgame which in September serves so well to fill the gap, have by the end of the month been pretty well thinned; moreover the survivors, to a considerable extent, leave the high moors during the "stubble period" to feed on grain in the lower lands, and as those which remain are now hopelessly wild, the season for shooting Blackgame over dogs *per se*, is thus practically over before mid-October.

But in October a new era opens in the grouse-shooting; so that, between the two game-birds, there is no lack of employment for the gun, and no break in the sequence of sport on the moors.

This new era arises from the altered habits of the grouse. They no longer cower in packs among the shaggy heather of some inaccessible slope, where they were wont to shelter from the bleak rains of September. The first bright frosty mornings in October rapidly dissolve the packs in twos, fours, or sixes, scattered all over the hills, and sitting bold and conspicuous on every knowe. At this season a man with a good eye, and who knows where to look for them, may get several brace in a day on rugged or broken ground, either by stalking the birds he has viewed sitting, or by "edging" them from the gullies or peat-ravines which intersect most moors. In the latter case excellent point-shots may be obtained by the assistance of an old dog which understands the business, and will indicate the position of (unseen) birds on the ground above, but which must never show up, itself, on the sky-line.

The number of grouse which can be killed thus is often considerable, for, not only can advantage be taken of ravines or water-courses, but every crag, and in fact every inequality of the ground, if of sufficient abruptness, will serve to conceal approach. Thus a thorough knowledge of the exact lie of the land, and the relation of gradients, &c., will enable its possessor to approach birds which might otherwise appear

"SUSPICION."

quite inaccessible, and to obtain the maximum of shots, while disturbing the minimum of ground. Still, there are on all moors wide stretches of flat ground, wholly devoid of any "advantage" to the gunner, on which he may often see dotted about some dozens of his handsome quarry, daintily strutting on the short heather, or loudly "becc-ing" out their defiant challenge, as though in conscious security. Of course a few shots may be got at these birds by short drives, and a brace or two may be killed. But there is a method of dealing with them in a more comprehensive style, which is often practised on the Border moors, and which I think is sufficiently interesting to merit a description. This method is to out-manœuvre the grouse with a horse and cart. There is of course nothing new in the idea. The stalking horse was one of the first inventions of the aucipial mind in long past ages, and to this day is used in many lands to approach wildfowl. Within my own experience, trained ponies are regularly employed in the "marismas" of southern Spain to gain access to wild geese and other fowl; and in the

"CHALLENGE."

same country even the Great Bustard is shot from a farm cart when leading the corn off the stubbles in July.

To return to the grouse: the main drawback to the system of "carting," and one which makes one reluctant to say anything about it, is the opportunity which (especially on certain days) it affords to the pot-hunter to seize a most unfair and unsportsmanlike advantage of the game. However, as this poaching, loafing, "half-a-crown-a-brace" sort of gentry unfortunately know already quite as much as I can

"CONFIDENCE." (CARTING.)

To face page 112.

tell them, I will assume that I am writing only for those whose object is sport, pure and simple.

A fine frosty October morning usually finds the Grouse scattered about in pairs or in small parties. On taking a cart on to the moors, it will soon be observed that their conduct is different to what would have been the case had a human being appeared alone, when they would at once have risen, perhaps a quarter of a mile away, and flown right out of sight. Of the lumbering, creaking cart they evince less suspicion. On approaching a pair (if the "helmsman" manages aright, and does not lead too directly upon the birds, rather circling round as though about to pass by them, while in reality drawing nearer every moment) the Grouse will probably show but little signs of alarm, and the gunner presently find himself almost within shot. The cock Grouse boldly sits erect, or slowly struts a little further away, while his crouching mate is visible a yard or two beyond, creeping low, and far less conspicuous, through the rough grass or heather. If the birds are restless, and do not at once, or at the first attempt, allow an advance to the fatal range, still the use of the cart gives this great advantage, that it does not alarm them in anything like the degree that the appearance of a man alone would have done. The Grouse at first may be suspicious or restless, and several preliminary attempts may result in failure; but their flights will be short; they only go, perhaps, a couple of hundred yards, when the cock suddenly flings himself some 15 or 20 yards almost vertically into the air, poises for an instant, and pitches straight down with a loud "bec-bec-bec"—adding, as soon as he is down, "c'm back! c'm back! c'm back!" Then his partner joins him, and soon by a little patience, and skilful leading, the sportsman succeeds in attaining the deadly distance, when the pair will offer as fine a double rise as can be desired even in August.

Where there are more than two birds, the chances are that they will require rather more manœuvring. The larger their numbers, the greater the probability of there being a restless spirit or two among them, which will shift them all.

But on a favourable day, this is no reason to give up the

effort. Sooner or later patient strategy will outwit them, and then perhaps a dozen Grouse may be seen, to the amazement of a new hand, carelessly sitting or running about within fair range of his gun. Few phases of ornithic instinct are more remarkable than the spell which this simple stratagem exerts on so wild a bird as the Grouse. When after successfully approaching a pack on the open moor, they at last rise at perhaps some five-and-twenty or thirty yards and a brace fall to the gun, one would suppose that then, at least, they would realize the danger. But it is not so. The charm remains unbroken so long as the gunner remains close alongside the cart. Once let his figure appear in separate outline and it ceases. The very fact of one or more of their number having fallen, appears to act as an antidote to the suspicions of the rest, and the survivors will perhaps pitch again close beyond their fallen companions. These latter, so the unharmed Grouse appear to reason, have only lit again, and the noise of the gun being partly drowned by the rumbling and creaking of the cart-wheels, tends to add to their delusion. Incredible as such hallucination may appear in so highly developed a bird-form, I can state from my own experience that three or four brace of wild November Grouse are not infrequently obtained within a space of a few minutes, and a radius of less than 100 yards.

It should, however, be mentioned that the manner in which Grouse receive the cart varies greatly on different days. There is no hard-and-fast rule, but, speaking generally, they "cart" best in fine, sharp, frosty weather, with bright sun and little wind. In rough, wild, or wet days they become packed, and as a rule, inaccessible. But there is a wide range in their individual dispositions. Some are always shy; on the very best of days, in the words of an old keeper, "there's always some 'll cart none!" and, on the other hand, one is sometimes agreeably surprised to get a fair bag on what appears the very worst of days.

The disposition of Grouse to "cart" also varies in different moors, or parts of a moor. As a rule they "cart" best on the whiter ground, where bent grass, rushes, and "spratt" are mixed with the heather, and of level or undulating contour. On high-lying black ground, it is difficult from the nature of the hills to take a cart, and such places are hardly worth the risk and labour of trying, as the birds will seldom "cart" at all there—that is on the heights, for they often do so very well on *lower* black ground apparently much the same in its physical features. Mixed ground is always the best, provided it is not choked with bracken, which prevents one seeing the birds; but the capabilities of any given locality can only be ascertained from actual practice.

The *modus operandi* is as follows:—The horse is led by the keeper, the gun keeping close in behind him—*i.e.*, alongside the horse's near shoulder. Should there be two guns, the best position for the second is at the tail of the cart, whence he can shoot on either side. A single gun, however, is preferable, as I will shortly show. When Grouse are descried, both men should keep close in, and avoid presenting a separate outline to the bird's view, and success is largely dependent on the skilful handling of the horse. The driver should not lead too directly upon the birds—rather circling round them, and without appearing to notice them at all. Of course not a word must then be spoken, or any signs or conspicuous movements made, and it is imperative never to stop the cart or to leave the horse's side—say to pick up dead, or any other purpose. I repeat this to emphasize its importance. Dead birds must be gathered from alongside, and the approximate position of winged ones may be marked by dropping a white handkerchief, so long as fresh birds may be in sight—however distant. The runners can be picked up afterwards (*i.e.*, as soon as operations in hand are concluded) by loosing the dog—an old setter is the best—which is tethered under the cart. A setter stands the "weather" best, and an old dog is preferable to a young one, inasmuch as the latter fights and "tews" itself on the chain till half suffocated, and panting like a grampus. The sagacity developed in an old dog when accustomed to this work is surprising, and one of the charms of the sport. Immediately on being let go, he casts away, alone, to leeward, and one after another "spots" the cripples. One may still (provided any fresh birds remain in sight) find it desirable to cover one's approach with the cart, and in fact go up to the points with the whole cumbrous paraphernalia. Dogs get quite enamoured of this sport, appearing to realize intuitively the whole of the strategy, and it is most amusing to see an old hand, which from long habit, persists in running *under* a carriage or dog cart, even along the high road.

A word now as to the horse; very few will at first stand having a gun fired so near their heads. Some never become accustomed to it at all, breaking into a canter at every shot, which is awkward on rough ground. An animal must therefore be selected which is capable, by his docile nature, of learning the lesson; such precautions as stuffing his ears with tow, which deadens the sound, may at first be adopted. After a little experience, a docile animal will pay no attention whatever to the shooting. But it should be laid down as an unalterable rule *never to fire across his head.* Only a few extra shots could be obtained thus, and it is almost fatal to the steadiness of any horse, and especially of a nervous one. Two or three such cross-shots will completely ruin many a horse for the work.

The degree of equine intuition of the sport which is developed, is only second to the canine already mentioned. I have seen a horse which had been terrified by a few reckless cross-shots, become so acute at seeing the Grouse, that he would sometimes even detect them before we did so ourselves, and then obstinately refuse to advance another yard, backing and rearing in complete terror—of course fatal to the object in view. I shall never forget one afternoon, when having borrowed a very old horse from a farmer, at the first shot the animal broke into a terrified trot, and for a full hour kept us running hard across and across a rough bit of fell—the while making frantic, but utterly futile efforts to pull him up. At last we got him on to a road, along which he continued his mad career, and here I left him to the keeper's charge, for during the struggle I had had to throw away my gun, and it cost me near a couple of hours hunting before I found it again.

Thus it will be seen that both horse and dog may present

some preliminary difficulties; but a greater and less easily mastered one to a beginner is *to see the birds*. Supposing him to be well equipped with horse, cart, and dog, still he may travel far and wide over fell and flowe without being able to see a single bird within shot. Probably after an hour or two's monotonous tramping alongside the cart, he will then begin to think the whole affair a delusion, and will leave the cart in disgust. The moment he does so, he will perhaps see Grouse after Grouse rise from the very heather he has just come through, and which he has scrutinized so closely without seeing anything in it. I have already referred (see p. 102) to the remarkable inability at first experienced in detecting Grouse when sitting low among the heather, and to the equally remarkable acuteness of vision which is developed by long practice.

It is obviously no easy undertaking to manœuvre a horse and cart on such rugged and treacherous ground as the moors. It would be the height of folly to attempt it, unless well acquainted with the locality, or accompanied by a man who is. Even then falls and minor accidents will occur, but it is surprising how much can be done, and what inaccessible-looking spots reached by a good man. Not only, therefore, should the keeper, or leader, possess a good eye, but he must also be intimately acquainted with every natural feature of the ground operated upon. Every bog or moss-hag, every gully or burn—he should know their positions and extent to a yard. His knowledge of the ground should, in fact, be as accurate as that of a "mud pilot" navigating a long P. and O. steamer amidst the tortuous channels and shifting sand-banks of the Thames estuary. Otherwise the result is not difficult to foretell. Even if by good luck he manages to avoid being bogged, or "couping" the cart over a boulder, or in a hidden drain, he will at any rate constantly find himself on the wrong side of some deep gully or impassable burn, with no ford perhaps for half a mile, and in which direction he will be helplessly ignorant.

As before mentioned, a single gun is quite as effective, or even more so, with the cart than two—any larger number is quite out of the question. A single man can concentrate undivided attention on watching his opportunity, and if possessed of a moderate amount of judgment, should at once decide when birds rise, whether to fire or not. Where two guns are shooting together, it frequently happens that bungles occur through misunderstanding, and advantages are lost through divided counsels or faltering decision. Though long shots, as a rule, should be avoided, it sometimes happens after repeated failures to approach a pack (these being invariably wilder than smaller numbers), that if a single bird be killed at a long distance, the remainder become more negotiable. They perhaps alight all scattered in a long broken line, just beyond the shot bird, and two or three brace may result which would not otherwise have been secured.

From what I have written, it will be seen that it constantly happens that the sportsman has the opportunity of firing at Grouse which are still sitting on the ground. It is needless to say that such shots should never be taken. They are a breach of the canons of fair play, and cannot be too strongly deprecated. Not only are such shots repulsive to sportsman-like instinct, and entirely eliminate one of the chief pleasures of shooting—*i.e.*, the handling of the gun—but as a matter of mere "blood and feathers" they do not "pay." So strong a bird as the Grouse, clad in his full steely plumage of winter, is much less vulnerable when sitting than on the wing. Unless struck on the head or neck, or on the exposed point of the wing (the rest being buried under the flank feathers), a Grouse will nearly always manage to rise, however hard hit, and fly a considerable distance, probably getting out of sight. This of course necessitates hunting a dog to recover him, thereby wasting much time and disturbing a wide extent of ground. Still there are people who from excitement, or (infinitely worse) from sheer greed, seem unable to resist the sitting shots; and it was to the *latter* class I referred before, who on certain days, when the Grouse appear quite lethargic, can perhaps massacre a dozen brace without making a single fair shot. The dastardly advantage which is thus afforded to the unscrupulous, is a serious drawback to the use of the cart by pot-hunters, who on the same principle would blow up a hare on her seat, or pheasants running forward on a drive.

Perhaps the greatest charm of this branch of moorland sport, is the frequent opportunity it affords of observing various wild birds close at hand and yet unconcerned. It is seldom indeed that one can enter so completely, as it were, into the privacy and domestic life of wild creatures. In August, one only sees the Grouse spring from their heathery refuge and spin away in fright; and on a drive the acquaintance is even shorter. Swish! and they are gone—a string of brown crescents dipping over the ridge. But alongside a cart it is different—perhaps only in punt-gunning can such chances be enjoyed of deliberately watching, close at hand, the ways and customs, the postures and contour of creatures to whom the human presence is anathema. In the one case they are careless, in the other unconscious of the hated presence. In both pursuits it is not too much to say that to any one who, like the author, has a love of natural observation as keen as the love of sport, the pleasure is doubled, or trebled. On a fine dry day, as one draws near, some of the Grouse will be observed to be lying down, basking in the sunshine, and in various postures—one perhaps resting on his side with one wing and one leg fully extended; others slowly loiter about the black ground, picking up odd bits of gravel to aid digestion, or a grass seed, by way ot pastime. They do not trouble the heather during the day, for Grouse only feed in earnest once a day—that is towards evening. Then one observes little amatory skirmishes and reconciliations—this even as early as October, a phase in Grouse life to which I have referred elsewhere. (*See* pp. 91 and 102). If the ground is wet with rain overnight, the Grouse avoid long heather or grass, seeking the barest places, where the old cock walks about with his tail carried nearly as high as his head to keep it dry. During rain they will be found huddled up into round balls of feathers, sitting on wall-tops, on stones, or any slight elevation where they can keep their feet dry. On such days, as a rule, they are utterly silent, whereas on a fine frosty morning the moor rings with their calls, and a cheery sound they are to the gunner as he enters on the moor at daybreak, and the cart wheels crush and grind over the frozen ground.

The different manner in which various species of birds regard this stratagem, and the varied systems of reasoning by which they appear to be respectively actuated in avoiding it, or otherwise, are also an interesting study. In a series of years' experience of "carting," one falls in with many kinds of birds, and their respective gradations of instinctive or reasoning power are curious and interesting. The Grouse, which certainly has the most highly organized mental faculties, seem to have reasoned out the whole problem. The cart they regard as an ordinary implement of every-day farm service; they frequently see carts in September and October crossing the moors to collect peats or brackens (which are used locally in place of straw), or passing along the open roads which traverse the fells. Thus they dis-sociate its appearance from the class of human beings which they have learned to hold in dread. It is a further illustration of their acute reasoning capacities that they discriminate between a shepherd with his stick and his colley, and a solitary sportsman with gun and setter, though the two apparently differ but little in general appearance. Instances in clear proof of this have occurred under my observation. After marking some Grouse down at a distance, and while reconnoitring the ground before beginning an advance, a shepherd has suddenly appeared on the scene. Unconscious of the proximity of man or birds, he passes with his lish gait and long swinging stride almost within gunshot of the watchful Grouse. They know him well enough; "it's only the shepherd!" and they cower down in the heather till he and his dog are gone by. But attempt to go and do the like! They are gone ere one's nose is fairly clear of the sky-line.

The reason which actuates the survivors of a small party to stop, after some of their number have just fallen to the gun, has already been alluded to—a remarkable fact, though it is known that even such wary fowl as Wigeon will occasionally do the like, alighting again among the slain after a shot from a punt-gun. This is always at night. Plovers

FIRST SIGNS OF ALARM. (BLACKGAME.)

To face page 120.

and other small shore-birds also wheel back, and even re-settle under similar circumstances, but in their case curiosity, sympathy, or simply ignorance appears to be the impelling motive, and not a distinct and intelligible line of deliberate reasoning (albeit based upon a mistaken premise), as is the case with Grouse. It sometimes happens that a Grouse (most often a single bird), on finding that the cart with its human adjuncts have approached nearer than he intended or cares about, will, instead of flying, squat or "scrogg" as it is called, among the heather. Nothing will then induce him to rise, and after circling round the spot, one sees the remarkable sight of a wild November Grouse, in full power of wing, lying motionless (but still apparently fearless) almost at one's feet.* I say fearless, for when he eventually rises, he will do so boldly and with a loud defiant "bec-bec-bec," as though his only feeling was one of annoyance at having been discovered—very different from the low and terror-stricken dash with which a *wounded* bird (which *has* realized the danger) will spring away from a close point.

Partridges are often fallen in with on the outskirts of the moor, and *if seen*, they will usually "cart"; but as they do not show, preferring to run like rats among the rough grass, one seldom or never troubles them, and I only mention them here as illustrative of their habitual instinct in this respect. Moor Partridge, moreover, never get really wild, as their congeners of the stubbles do, for there is always plenty of cover. The lack of this desideratum in this era of shaving reapers and close cut hedgerows, is one of the chief reasons for Partridge ever becoming wild at all. On the moors, when broken, they will lie to points all through the season.

Next to the Grouse, our most important game-bird on the Border moors is the Blackcock. One sees them every day in packs. They will not "cart" at all. That is the rule; but it has its exceptions. Thus one occasionally finds oneself close up to the Grey-hens. But they are not very wild birds; they are emboldened by being so frequently spared. As this kind of game does not pair, the Grey-hens

* An old cock Pheasant will sometimes do the same, on a bare stubble or fallow, in hopes of being undiscovered.

are habitually "let off," and thus it happens that from carelessness, simplicity, or confidence, they sometimes disregard the proximity of a cart. That is the conclusion which the manners of Grey-hens when close at hand, seem to indicate. But the case is very different with the Blackcock. He is neither simple nor careless, nor apt to trust. *Timeo Danaos* is his motto, and he will never "cart." Only on four or five occasions have I managed to get an old Blackcock from the cart, and these, with one exception, merely by chance shots at long distances. The exception was a singular instance of the erratic properties of bird-instinct. We were on a hill alive with Grouse, and though it was a fine bright December morning, with keen black frost, not a bird would "cart." At last we observed a single Blackcock, which not only allowed us to reach an easy range, but actually, at that moment, squatted flat on the ground. Such tactics were so extremely unusual in a Blackcock, that we continued advancing in order to see what he meant to do; prostrate and motionless, he lay till we were actually within three yards, when he sprang away on his last short journey.

On approaching Blackcocks with a cart, they are all visibly on the alert, with necks up on full stretch, and evidently unable to comprehend the *raison d'être* of the phenomenon. It may of course be only a harmless farm-cart; but if ever they commence so to grapple with the problem, their distrust and suspicion invariably overcome their reasoning powers, and they take wing at three or four gunshots' distance.

The entire failure of this stratagem to outwit these birds arises, therefore, from no superior intellectual development in the Blackcock as compared with the Grouse, but really the reverse. Intellectually, *Tetrao tetrix* is distinctly of an inferior capacity to *Lagopus scoticus*—cunning, distrust and suspicion predominating, rather than reason or calculating power. It is, in short, rather in despite than by reason of his mental capacity that he manages to escape.

Yet, although Blackcocks cannot be approached directly with a cart, it frequently happens that some are bagged during a day's carting. Being so conspicuous an object, their position can generally be detected far away, and by manœuvring the cart at a cautious distance (*i.e.*, just so far off as will not put them up) a position may often be attained, whence an advantage is secured for a stalk. The birds will then continue to watch the harmless cart, while the gunner, having slipped away from its side, is creeping on them under shelter of a gully or dip in the ground—even a patch of tall rushes is sometimes sufficient.

Golden Plovers are among the most erratic of birds in their disposition. Some days they are so careless one can walk almost openly within shot, or, with a cart, lead round

RISING TO CLEAR THE DYKE.

a pack sitting listless and plunged apparently in deepest meditation, till one gets the greatest possible number into line. On other days they will hardly remain in the same parish with a human being. I remember one day vainly endeavouring to outmanœuvre a pack of about 100 on an open flowe. First we tried the cart, then stalking and driving—all to no purpose. Eventually I left them, and was walking away, when I heard behind me a confused chorus of pipes. The whole pack were wheeling past within thirty yards ! The advantage they had so persistently denied to a series of efforts, they now placed spontaneously in my hands, and five of their number forfeited their lives to that erratic behaviour.

None of the true wildfowl (*i.e.*, Ducks and Geese) can be

outwitted thus, at least in this country. They are too utterly afraid of man and all his works. Over two centuries ago, Gervase Markham wrote (in 1655): "A fowle is wonderfully fearful of a man," and they certainly are not less so at the present day. The Grey Geese, which appear crossing the moors in October, will pass right overhead rather than alter their course; but they are always at a height which they consider beyond all reach of danger, and I never knew them miscalculate. The Curlew is one of the most watchful and suspicious of birds; but though we do not have them on the moors in autumn, I feel sure that, if there, they would thus fall an occasional victim to misplaced sagacity, having frequently passed within 100 yards of them when driving across the sand-flats of the coast without their taking any notice or even discontinuing their probing operations.* I have, however, perhaps dilated too much on this subject of bird-instincts, which has been to me a most interesting study.

The foregoing is a rough outline of a system of sport of which it is sometimes useful to possess some knowledge late in the season. Many moors are too irregular in shape or in contour, or too narrow to admit of driving, and yet may perhaps be too level to afford the slightest chance of approaching wild birds. On such grounds, "carting" offers a means of obtaining access to them, and six, eight, or twelve brace may thus be killed, where hardly a bird would reward the utmost labour otherwise. Driving also necessitates at least four or five guns, and a larger number of drivers—conditions not always attainable. A single sportsman may not only, under favourable conditions, enjoy excellent sport in this way, but he will be delighted with the constant and interesting opportunities for the observation of wild birds.

* Since writing the above, I have proved that this supposition is correct. As an experiment, I took a cart to-day (Jan. 27) several times within shot of Curlews, both single birds, and small flocks, on some wide sand-flats on the coast. Even when a bird was shot, the rest only flew a short distance, pitching again within a quarter of a mile; but were considerably more suspicious on a second approach.

BIRD-LIFE ON THE MOORS IN NOVEMBER.

With November usually comes winter, and always the Snow-Buntings, charming little birds apparently ordained by Nature ever to enliven the most dreary prospects of snow and ice. Cruel as such a destiny may appear, Nature has evidently provided that it should not be so to them, for no bird appears more blythe and joyous than the Snow-fleck. Whether one sees him in summer, clad in black and white (as the writer has done), among the grim and forbidding lands of the Arctic Seas, or in winter on our own storm-swept hills, he is always the same—bright and happy, despite the most dreary surroundings, seeming full of life and exuberant spirits, flicking about more like a big butterfly than a bird; and his little triple trill is as cheery as his actions. It rather resembles the rattle of a Partridge's wing, and one looks round to see if some have passed behind. The Snow-flecks arrive on the moors with great regularity about the 1st of November, almost to a day, usually in large flocks, and feed on the seeds of grass and rushes. Wherever a single seed-bearing blade rises above the level of the snow, their little footsteps may be seen imprinted on its surface. These earlier flocks are almost entirely composed of immature birds. The adults, with their winter plumage, are comparatively scarce at all times, though they are rather more numerous in mid-winter.

In November the salmon leave the larger rivers and enter the burns, while the trout, which spawn earlier, push far up the tiniest hill streams, often taking advantage of a "spate" to reach spots high out on the fells, where the streamlet has dwindled to a mere drain, with the shaggy heather meeting over its waters. Of this fact the solitary Heron is well aware, and his great grey form is a characteristic feature of this season, solemnly flapping across the moors to some little burn that he wots of as a favourite resort of the trout; or perhaps he startles a nervous shooter by suddenly flapping out under his very feet from some deep-sided hidden little burnlet, where the sportsman would as soon expect to find a Dodo as either Heron or trout.

Another bird which seeks the higher ground in November is the Water-Ouzel or Dipper. Usually these little fellows, as well as the Heron, prefer to frequent the larger burns in the valleys, but at the end of October, and in November, it is a frequent occurrence to almost step upon a Dipper concealed in one of these little overgrown drains far up on the hills. Their object in going there has, *primâ facie*, a slightly suspicious appearance, but it has, I believe, been conclusively proved that the Dipper is almost, if not entirely, guiltless of devouring spawn. Even if they did commit so heinous an offence, the damage resultant would, in my opinion, be wholly imperceptible, at least in the northern streams; but *homo sapiens* is nowadays very intolerant—often unintelligently intolerant—of even the appearance of rivalry, in anything of which he may have arrogated to himself the monopoly.

I have noticed in November a Dipper swimming about, like a little duck, on the open water of a lough at one of the highest points on the moors. The Kitty Wren is, besides the Dipper, almost the only one of the resident small birds that one sees on the moors in November; and a strange little creature it is to meet with among the bare heather, far out on the hills, at this chill bleak season. Though generically so distinct, there is in several points a singular resemblance between these two little winter songsters. In form, carriage, song, and in their mode of nesting they are strangely alike, and agree moreover in their common hardihood and endurance of extreme severities of weather, both species braving the most protracted snow storms on the Border moors.

The Raven has now become comparatively a scarce bird on the Northumbrian side of the Borders; but a few still remain, and in the stormy days of February resort to their rock-fastnesses, and to the most remote and inaccessible crags among the hills, where they commence to nest while yet winter remains in full possession of the land. And in the autumn months the slow and powerful flight of the Raven is not an infrequent spectacle on the lonely moors; or perhaps its presence is only indicated by the single hoarse croak from a black speck far up in the clouds—sights and sounds which, by the way, to an eye appreciative of what is wild and fitting in Nature, are full of appropriateness and of the *genius loci*.

Though I have observed Ravens on the Northumberland moors in all the autumn months, the best view I ever had at close quarters was on November 4, 1882, when we came suddenly on a pair near Elsdon. With heavy, powerful flaps they were endeavouring to beat to windward against a hurricane from the west, flying low on the heather. Eventually they appeared compelled to give up the attempt, and drifted away to leeward, passing us again within a short distance. We afterwards found a dead sheep, minus his eyes, lying in the direction they had been making for. In the month of August, in 1884, Ravens were more numerous than I ever before observed. We had had, locally, a sharp attack of Grouse disease (in its sudden or virulent form), and the numbers of dead Grouse lying about the fells had perhaps attracted the Ravens from more fortunate moors at a distance. I saw a Raven, an adult female, shot on Hudspeth, Northumberland, September 6, 1879; its effigy is now in the Sunderland Museum.

During November, parties of small Gulls occasionally appear, flying across the moors—probably Black-headed Gulls paying an interim visit to their inland breeding-places, as Rooks assemble at the rookeries at the same season. The Lesser Black-backed Gull also appears now and then, and, with his wide stretch of wing and boldly-defined plumage, forms a strikingly conspicuous object in the moorland scene.

Another visitant characterestic of this season is the Short-eared Owl. They arrive in this country towards the end of October—the 22nd is the earliest date I have noticed one;

and, though their favourite resort is the stretch of bent-grown sand-links which fringe a great part of the coast-line, yet a sprinkling of them is distributed over the moorlands. By day these Owls usually sit somnolent among the beds of rank rushes or tussocks of "white grass." I have put them up in turnip fields. But on dull days and in rough weather they occasionally commence to hunt by daylight and, when seen thus, have a singularly hawk-like appearance, very like a rather pale-coloured female Hen-harrier. On the afternoon of November 1, 1881, during a blinding storm of driving snow, one of these Owls passed over us, soaring strongly in wide circles, but gradually drifting to leeward. Through the drift he loomed as big as a Buzzard, for which he might easily have been mistaken.

The quantities of mice which at this season are found far out on the open moors are, no doubt, the chief attraction to the Owls. The number of those little creatures one sometimes sees in the course of a day's shooting is surprising. I may add, in reference to this species, that I have observed what I have no doubt was a Short-eared Owl (hunting by day) on September 1, and another (obtained) as late as March 29, which dates make their nesting in this neighbourhood appear probable.

By the month of November the general autumnal migration (so far as the moorlands are concerned) appears to have been completed, and bird movements are, for the time, suspended. Most birds appear to have settled themselves down into their permanent winter quarters. Thus, for example, the following figures, showing the relative abundance of Snipe on the moors during the different autumnal months, prove that arrivals of that bird have ceased for the year. The figures, it should be stated, are *relative*, not actual; they are extracted from records of ten seasons and reduced proportionately to the number of shooting days in each month:

	Aug.	Sept.	Oct.	Nov.	Dec.
Snipes killed . .	90 .	125 .	75 .	30 .	25

The slightly reduced numbers in December are merely due to the greater prevalence of snow during that month. Snow drives the Snipe from the moors to the lowlands. Thus at

Silksworth, co. Durham (lowland), where I have shot for upwards of twenty years, we rarely see a Snipe during ordinary weather; but directly we have snow, down come the Snipe from the hills. By walking round the little open burns before breakfast on the morning after a snow-storm, I can always find several Snipe; and if these are killed others take their place by next day, and so on during the continuance of the snow. But on the first indication of a thaw away they go at once.

Snipe, as every one who has followed the pursuit of Snipe-shooting knows, are the most shifty and capricious of all game-birds—here to-day, gone to-morrow; sometimes sitting only on the highest and driest places, at others all congregated in the bogs or along the edges of moss-hags. Then the versatility of their "attitude" towards the fowler is equally noticeable, one day sitting close, while another they spring at a couple of gun-shots' distance. It is sometimes difficult to account for all their vagaries, though doubtless there is a reason for each change in their haunts and habits. The main factor, besides the weather, in influencing their movements appears to be the moon. Snipe and Woodcock being both night-feeding birds, it is, of course, well known to sportsmen that they are only exceptionally found on feed during the day. It is only the youthful enthusiast who splashes about in water nearly knee-deep to find either the one or the other. Both birds have a special antipathy to being wet, and, as a rule, by day sit high and dry, though no doubt plenty of water knee-deep may be found close at hand. As a general rule, it is safe to state that Snipe will lie best, and be found in the driest places, during the period of full moon, especially if the nights be fine and fair. They have then been able to feed abundantly all night, and consequently are more apt to be resting and to lie close during the day. But on dark nights when there is no moon, or in wild weather when the moon is overcast, they are compelled to feed partially by day, and at such times are more watchful and wild. I have occasionally had opportunities of observing Snipes while feeding, sometimes with their breasts quite half immersed as they probed about in the shallow water.

Woodcocks appear rather more addicted to completing their prandial operations during the hours of daylight—during snow I have frequently observed them feeding on the edges of small open burns; and, indeed, to the arcana of many of the deep bosky dells in which they delight, the weak horizontal rays of the winter sun hardly penetrate at all; and in such places they can feed almost undisturbed by daylight, even at noontide. I need hardly say the popular notion that Snipes and Woodcocks live by "suction" is a delusion, unless a good fat worm be included in that term.

The prevailing winds during autumn on the Northumbrian hills are west, or westerly; and these are usually dry

WOODCOCK.

winds. However threatening the heavens may appear, very little rain will fall so long as the wind's "airt" is westerly. East winds, on the contrary, are usually accompanied by rain, fog and dirty weather. A change of wind from east to west is frequently attended by an almost simultaneous clearing-up of the water-logged atmosphere, and shortly succeeded by the welcome appearance of the sun. The cloud effects produced on such occasions are often wonderfully fine. I well remember a magnificent spectacle displayed late one September afternoon. After several days' incessant rain, fog, and strong east wind, suddenly the wind fell, then shifted round by the south to west. In half an hour the rain ceased, the murky cloud masses were driven back upon each other—literally rolled together like a scroll, and hurled back

GONE TO ROOST.

upon the North Sea. Then the bright rays of the sun completed as sudden and as perfect a transformation scene as can be imagined. Under the effects of the genial change and bright sunshine the soaked and dripping heather soon began to dry, and along the topmost ridges the Grouse rapidly congregated to enjoy the unwonted warmth. It was about four o'clock, and my bag almost empty—indeed, for some days I had done next to nothing, under the adverse conditions of rain, fog, and mist thick as peasoup. But now the chance had come, and I hastened to take advantage of it; by working under the wind, and keeping constantly to the shaded eastern slopes of the ridges, I got shot after shot at the wild September Grouse, which sat close under the highest crests, revelling in the last rays of the declining sun.

In reference to the weather—though weather forecasts are notoriously unreliable (amateur ones at any rate), it is frequently the case that a very wet morning *early* is succeeded —say at 10 or 11 A.M.—by a fine bright day. Early on such mornings, as we have peered out into the still unbroken darkness, with the pitiless rain descending in "whole water," our sinking spirits have been cheered by our worthy hostess's remark, "It's raining too hard for a wet day!" a forecast which, in at least a clear majority of cases, has proved correct. On such occasions, therefore, I would advise the keenest sportsman to go quietly to bed again for a couple of hours, and try to comfort himself with the reflection that "good luck is better than early rising."

BIRD-LIFE ON THE MOORS IN DECEMBER.

THE month of December, in a mild season, does not materially differ, ornithologically speaking, from November, as described in the last chapter. The Grouse, still largely in pairs or in small lots chiefly composed of aggregated pairs, continue to hold their matutinal concerts, and the old Blackcocks bubble and sneeze on their chosen hillocks. Even the young cocks of the year now commence to strut and caper among the assembled Grey-hens, as I have already mentioned. Despite the thinning of their ranks by three or four months' shooting, both Grouse and Blackgame appear fully as numerous in December as at any previous period of the autumn, sometimes even more so. It seems probable, or even certain, that we have in Northumberland at this season a considerable immigration of both kinds (but especially of Blackgame), which have moved southward to escape the greater severities of winter on the higher and more exposed moorlands further north. With this exception, but little change is perceptible in the habits or distribution of the regular moor-frequenting birds in a mild December.

But December is not always mild. On the contrary, about one year in every three is characterized on the Border hills by severe and protracted snow-storms, which completely change the face of nature, and greatly alter the normal conditions of wild life. Than the appearance of the great rolling hills, when newly enveloped in their wintry mantle, there are few more imposing, and, *sui generis*, more lovely spectacles. Where in August one's eye ranged over a wide succession of hills and vales all purple with the bloom of the heather, and varied with the many bright tints which beautify the wild moorlands, now nothing is visible but a vast expanse of glistening snow, unbroken save where crag or scaur stands out bare and black, casting a deep blue shadow across the slopes, or where dark patches of a pine wood struggle through their frozen burden. These conditions rather suit the idiosyncrasies of the writer, who has, under them, enjoyed many a hard and pleasant day's tramp among the hills. Starting away just as the tardy daylight begins to break, the low pleasing carol of the Dipper strikes cheerily on the ear from the burn where he sits on a stone in mid-stream, piping merrily away. Following the course of the burn, we presently come on another pair of these hardy little fellows, busily diving under the fringe of fast ice which lines either bank; then popping up quite unconcernedly in the narrow centre channel of open water, to land on the ice-edge. Far differently do their neighbours the Water-hens regard the new conditions of life. They seem utterly dismayed at the loss of their accustomed hiding-places among the rushes and sedge, and splash about in a constant state of fright. Even that most skulking of all birds, the Water Rail, is now at fault in his consummate mastery of the art of hiding, and runs disconsolately about the frozen reed-beds, seeking in vain to conceal his dark form when everything is white. Except at such times as this, so retiring are the Rails, that one is hardly even aware of their existence on the burns. Now they seem stupefied by the changed conditions, and so loth to take wing, that in the heavy snow-storms of December, 1869, we ran one down and captured him alive. The Spotted Crake is a scarcer bird, and does not remain with us in winter. The only one I have seen killed in the northern counties was shot October 26, 1883.

The hill burns are a favourite resort of mine during deep snow, for the chances they offer of Snipe and Duck (Mallard and Golden-eye) driven in from the frozen loughs above. But in long-protracted snow-storms the Mallards, after a time, forsake the inland streams altogether, and betake themselves to the open coast, where at such times I have found them fall a comparatively easy prey to the punt-gunner, being rather less suspicious of the long, low, white craft than are their congeners of the salt water. The Golden-eyes, on the contrary, being good divers and feeding at the bottom, appear but little affected by any weather, and continue on the inland streams long after the native Mallards have taken their departure. Only once—viz., in the terribly severe winter of January, 1881—have I seen these ducks compelled to retire to the salt water for a living. Teal dislike extreme cold; during hard winters I have never met with them on the moors at this season, and very rarely on the coast. The Heron and Jack Snipe agree in but few respects, especially in size, but both are characteristic of severe weather—the Heron on the larger burns, the Jack on the tiniest little open rills, where, as they take very short flights, three or four of them will perhaps afford sport to an indifferent shot for half a day.

Of the resident small birds, if the winter be mild and open, a fair number of species continue to enliven the scrubby birch and alder woods along the fell-edges and in the valleys. Among these may be mentioned large flocks of Chaffinches, Greenfinches, and numerous small parties of Tits, while here and there a Goldcrest or a Tree-creeper may occasionally be observed. But it is only during mild weather that we have their company; when winter sets in in earnest they soon disappear. Hardly one of them remains after two or three days of snow. The Wren remains steadfast; he simply moves off the hills into the woods, but the rest are all gone. As regards hardihood, in the first rank of all small birds stand the Wren, the Robin, the Bullfinch, and the Blackbird. (Thrushes entirely disappear from the fells in winter.) The four species above named appear immovable, not retiring before even the most extreme severities of weather. The Blackbird, however, appears to be the most affected, and after a week or ten days' hard weather many are in a moribund condition. The cock Bullfinch, with his crimson breast and sharply-contrasted plumage, is a conspicuously handsome object in the wintry landscape. A pair or two are often noticeable perched on tall hemlocks or on thistles, and busily feeding on their seeds when everything is deeply buried in snow.

The Moor Partridge, a local race, varying from the ordinary bird in its slightly smaller size, and in some differences in the disposition of the colouring on the head and neck (having also, I consider, a more delicate flavour), fares very hardly during long-protracted storms. They endeavour to eke out a subsistence where the snow is soft or shallow, or where a strong-flowing spring has cleared a patch of green grass; but if the snow continues, they become miserably thin, and a long winter on the hills would almost kill them out, but for human assistance.

Partridges differ from Grouse in never becoming really wild, providing there is sufficient covert. Wildness in Partridges appears to be solely a matter of want of covert. In the moorland districts, where there is plenty of rough grass, &c., at all seasons, they can be "broken" and shot over dogs in December and January, almost as certainly as in September. That is, they can, under favourable conditions, be got to *lie* nearly as closely; but their speed of wing and "dash" in getting away is now very much greater. I speak, of course, only of shooting in open weather, for in snow no one should molest the little brown birds.

When, after a hard climb, one reaches the fell tops, it might be thought an easy matter to find the Grouse on the level snow, where the smallest object can be seen at a great distance, and often magnified by the rarefaction of the air till a Snow-Bunting looms as large as a Blackcock. But it sometimes happens that one scans for miles the wide expanse of glistening snow till one's eyes ache with the brilliant monotony of the millions of sparkling crystals—but without seeing a single bird. The reason of this is that, in such cases, the Grouse are deeply buried under the snow, as one presently finds on coming across a perfect network of burrows where a pack has been resting. These burrows are often two or three feet deep, sloping downwards, like rabbit-holes. A favourite resort for these sub-niveal operations is a steep bank where the heather is old and long, and where its stalks keep the snow loose underneath. In such places a sentry is sometimes left to keep watch outside, but as often as not the precaution is neglected, and the whole pack appears fast asleep.

When strong winds prevail, it generally happens that large patches of heather are swept clear of snow on the weather-slopes of the hills; and to such places the moor-game resort in very large packs, finding there both plenty of food and also immunity from their enemies; for the snow which has been blown off the ridges forms huge "wreaths," or drifts, of great depth, lower down the slopes, and which are quite impenetrable to man. Failing this resource, Grouse and Blackgame are able to find food where sheep have partially uncovered the heather. In these ways, and being of so hardy a nature, it seldom happens that Grouse suffer much inconvenience from the utmost severity of the weather. Their principal danger arises from their appearing to lose themselves from want of their usual landmarks.

It frequently happens during such weather that one sees numbers of Grouse perched upon trees, especially on thorns. This is contrary to their usual habit (although I have seen two or three instances of it in open weather), and seems strikingly to demonstrate their affinity with *Lagopus subalpina*, the Rypa, or Willow-Grouse of Norway, which is no doubt the parent race (*cf.* Wheelwright's "Spring and Summer in Lapland," p. 334). Probably if we were visited in England by a period of severe winters (say, a few centuries), our Grouse would revert to this, the original form, acquiring the pure white breast (some approach to it now) and white primaries of *L. subalpina*, and would habitually prefer perching on trees to sitting among the heather.

The big Tawny or Wood-Owls are very noisy at this season just after dusk, and their loud demoniacal laughter, as a pair or two call and answer each other from the woods across a moorland valley, is singularly weird and uncanny. They sit by day in an ivied tree, or on an old Cushat's nest, and find abundance of prey by night in the mice which after dark scamper in all directions across the snow.

I will now conclude these notes on the moorland birds with the following table, showing the results of my last ten shooting days of the season (alone), in two consecutive years in Northumberland :—

	Last ten days. Mild season.	Last ten days. Deep snow on last three.
Grouse	101	119
Blackgame	10	17
Partridge	8	2
Pheasant	1	0
Golden Plover	5	7
Snipe	9	5
Mallard	0	1
Golden-eye	2	2
Heron	0	1
Hares	2	0
Rabbits	1	2
Total	139	156

This list may be interesting as showing both the variety of game, and also what can be done by hard work, even at the extreme end of the season, and in the short, dark days of November and December. It is open, I fear, to being thought egotistical; possibly it is a little so; but perhaps no one will be so unkind as to notice that.

WOOD-PIGEONS.

Of all the many fowls of the air which have contributed to the "cacoëthes cædendi" of the writer, there are few to which he owes a deeper debt of gratitude than to the homely Wood-Pigeon. The Wood-Pigeon, or Cushat as it is generally called in the north of England, comes every winter in throngs to our woods, and affords during the earlier months of the year almost as good sport as any bird that flies.

After many years of observation of their habits, it is yet impossible to lay down any general rules which would apply to their movements. They are among the most uncertain of birds. Like most of the true wildfowl (to which their habits bear some similarity), Cushats are here to-day, gone to-morrow; abundant one year or one month, scarce the next. Why, one can assign no reason, for, if we build up a theory, perhaps the very next year will completely refute it. To set down what appears to be as nearly an average as possible, there occurs one main annual immigration of these birds in the north of England some time about Christmas—occasionally this takes place as early as November, but sometimes not till the end of January. These new-comers are easily distinguishable, being conspicuously cleaner in their plumage, and lighter-coloured than the more grimy residents of the northern counties. Indeed, at that season, we have but few Cushats left of native breed, most of these departing (presumably southwards) as soon as their latest broods are fully fledged—*i.e.*, about September and October. There thus occurs an interval between their departure and the arrival of the great influx which takes place later in the season.

These newly-arrived, bright-plumaged birds are usually set

down as coming "from foreign," and there is no doubt that in some seasons, and under certain conditions of weather on the Continent, very great numbers of Wood-Pigeons do cross the North Sea, especially in the month of November. But, as it does not appear that any very great quantity of Cushats are bred in Norway and Sweden, or even in Denmark, it is probable that the migrations of many of our visitors are less extensive, and that they have only come from the Scottish Highlands, the Lothians, or other parts of these islands, merely shifting about in search of food requirements.

Then, after they have come, it is equally difficult to diagnose their movements, so restless and uncertain they always appear. Roughly speaking, they are usually most numerous in this neighbourhood (co. Durham) during severe weather; the more prolonged the winter, the more Cushats appear. But they are the slaves of the weather, and every change affects their numbers. Thus, though heavy snow may bring hundreds when few or none were here previously, yet in the very reverse, should there chance to be none during the hard weather, they will appear in thousands on the thaw.

All the day they spend on their feeding grounds among the turnip fields, stubbles, or clover lea, alternately feeding and taking a digestive siesta on the nearest hedge-trees; the birds in the latter position also acting as sentries, whether purposely or merely by accident.

A big pack of Cushats on the feed is easily made out a long way off by the habit of the rearmost birds of the flock continually flying up over their companions and alighting in the front rank, thus causing a constant movement. But it is a great mistake to molest them during the day; very few comparatively can ever be killed then. Whoever would fill his bag with Wood-Pigeons should leave them to get their feed in peace, and wait for them towards night in the woods where they go to roost. There, during the last hour of daylight, is the time and place, in a favourable season, to kill them by wholesale. As flight after flight pours in in quick succession, the gunner will be rewarded for his well-timed moderation during the day. But a very little shooting makes

Cushats excessively cautious, and they then circle high round the wood, wheeling perhaps three or four times before dropping within shot. This is a critical time for the gunner, standing grey and rigid against a tree, which he should resemble as closely as though he formed a component part. Scores of pairs of eyes keen as hawks' are wheeling overhead, searching the ground, and any movement or conspicuous colour will assuredly betray him.

There are two points in particular which are specially noticeable by the scrutinizing eyes overhead—namely, the white colour of an upturned human face, and the inward movement of the elbows on raising the gun to fire. The sharp-eyed birds instantly detect the change of outline in the "tree," and what would have been a fair incoming shot is instantly exchanged for an all but impossible "snap" end-on through the branches. Both dangers may be avoided by raising the gun vertically before one's face just before the Pigeons begin to pitch in. It is no use turning round to watch the birds as they circle behind one. If they have gone on, you are done with them; if not, and they continue wheeling round, one's eye picks them up as soon as they re-enter its circle of vision. As long as the loud "swish-swish" of their strong pinions is audible, the Pigeons are, as a rule, too high—no greater mistake than "gliffing" them by shots at impossible heights. The finest shots are when they are "pitching in," lowering their flight to alight, and then it is noiseless. When they alight within shot, it is a mistake to hurry to fire. They will not see a man, though full in view, *provided always* he remains rigid as death, with his back glued to the tree. Give them thirty seconds, then you can safely look round to see where two or more are sitting together or in line; or, by giving them two or three minutes, some may possibly begin to crowd, and a family shot is the result of a little patience. If two Pigeons are sitting a foot or two apart, it is just possible to kill them both sitting, by a very quick right and left—*i.e.*, after killing No. 1, the gun is instantaneously shifted on to the assumed position of No. 2. Of course, this is sharp work—it is an affair of fractions of a second—and if success is by no means assured, its occasional attainment is, for that reason, all the more gratifying.

One habit of the Cushats is fixed and invariable—they fly head to wind; consequently they may always be looked for to leeward. This habit they observe equally in the strongest and lightest breezes. So long as there is the slightest current of air to indicate the direction of the wind, from the opposite quarter the Pigeons will certainly approach, and, after their preliminary circling flight, will always finally alight in that direction. The position to select, therefore, for shooting them at flight is on the leeward side of the wood, about a gun-shot from the outside, and, if possible, opposite to the highest trees which may happen to grow there, and towards which they are pretty sure to direct their flight. Not only are the prettiest shots thus obtained at the Pigeons as they pitch downwards, but should they get past, and alight behind one, they are not so difficult of approach from that direction, especially in a strong wind. Indeed, a good breeze is a *sine quâ non* for a thoroughly successful night at Wood-Pigeons. On dead calm nights they are liable to drop in from all directions, and, however vigilant a watch is kept, some chances are sure to be lost through birds suddenly pitching in from behind. Besides which, on such still nights, it is hardly possible to move a yard without disturbing them—the least crack of a dead twig, or rustle of fallen leaves under one's feet, and they are gone.

The above remarks as to choice of position are, of course, only applicable to places previously untried; and much always depends on the shape and lie of the woods and other local conditions. But the sylvan geography soon becomes well known to a regular shooter, who in a few evenings ascertains the most advantageous spots and the favourite roosting-places of the Pigeons. One of the best woods in my knowledge is a small square clump of tall beeches, perhaps an acre in extent, and in a somewhat exposed position on rising ground. In this small wood hundreds of Pigeons have been killed in the season—say from December to March—the best evening's work being twenty-three birds, all single shots, to one gun. But during rough, stormy weather the Cushats avoid these tall, exposed beeches, and on such evenings the best sport is obtainable in lower-lying woods where the black Scotch firs abound, and which afford admirable shelter for the Pigeons, and also concealment for their enemy. On exceptional occasions, during very wild weather, with strong winds and snow, Pigeons fly very low, almost brushing the tree tops in their struggle to windward, and on such nights are easily obtained—the more so as one can move freely about (provided the underwood is tolerably clear), without much fear of disturbing them. But, although such chance occasions afford an opportunity of killing a considerable

WOOD-PIGEONS.—EVENING.

number, yet the sport is incomparably inferior to that in more moderate weather, when the Pigeons fly higher and more boldly, offering the greatest variety of shots, together with left-barrel chances of every degree of difficulty—up to the impossible. Where there are several woods to which they resort, a gun placed in each wood keeps them moving about; but it is a more deadly plan (for a single gun) to send a couple of boys round to tap the trees with a stick, this being quite sufficient to move the birds, but not to scare them right away.

During heavy snow, especially when the wind has covered the tree-trunks with the drift, it is often difficult to get sufficiently concealed. I remember one night in January, some years ago, being greatly disappointed through this circumstance. The Pigeons almost invariably detected me too soon, and, though there were many hundreds of them on flight, I only managed to get eight. A few days afterwards, under similar conditions, I tried the experiment of putting on a common white nightshirt over all, and a white flannel punting cap. This succeeded admirably, and that evening I got twenty-one out of a much smaller number than were seen on the previous occasion.

There is a sort of charm in the stillness of the wintry woods as the daylight fades away, and the gloom gradually deepens among the bare trunks—hardly a sign of life except little parties of Chaffinches and Titmice flitting about among the leafless branches, or the rustle of a mouse among the dead leaves at one's feet. Presently a Grey-backed Crow approaches with his triple croak. No bird in creation is sharper of eye, and it is almost ludicrous to see the aerial somersault he turns, when he discovers an ambush, and a pair of barrels rise right under him. Then the silence is broken by the call of the old cock Partridge on a stubble outside; calling together the scant remains of his once big brood, and as the darkness settles down, the low, cat-like whistle of the Long-eared Owl is a safe warning that it is time to gather up the spoils and be off home.

As winter commences to merge into spring, we once more get clean, bright-plumaged birds, evidently new arrivals from less smoky regions. From this and other reasons it is clear that a considerable movement of the pigeon-tribe takes place in early spring, usually in March. The winter stock, in all probability, return whence they came, and their places are taken by those which, having passed the winter further south, now return to the north to breed. By the beginning of April the Cushats commence nesting, and, though they only lay two eggs, make up for any shortcomings in that direction by continuing to rear successive broods over nearly half the year —for young birds remain unfledged in their nests till September, and often well into October. The home-bred Cushats at that period depart southwards, which brings me back to the date at which I commenced this sketch of their annual movements.

Of late years we have had a pleasing addition to the variety of Wood-pigeon shooting in the appearance, in the north of England, of his smaller relation, the Stock-dove. Few facts in local ornithology are more remarkable than the rapid extension of the northward range of this species, which formerly was totally unknown here. Out of some thousands of Wood-pigeons killed at Silksworth during the last twenty winters, there was not a single Stock-dove until four or five seasons back, when we got one, which was regarded as quite a rarity. The following winter three or four were obtained, but in the winter of 1884–5 they became quite numerous. Of fifty-three wild pigeons shot at Christmas-time no less than five were Stock-doves, and many others were with them. During the subsequent months of January and February, 1885, we obtained them quite commonly—indeed, two or three Stock-doves were killed almost every evening we went out to shoot Cushats, and they came to be regarded quite as a regular component part of the bag. They sometimes flew to roost in company with the Cushats, for birds of both species were once or twice killed out of the same flock; but more often the Stock-doves came in separately, in small parties of six or eight. They were easily distinguishable from Cushats when on the wing, by their more rapid, impetuous flight, as well as by their much smaller size, as the following weights will show: Wood-pigeons, weight 17oz. to 26oz.; Stock-doves, weight 12¾oz. to 14½oz.

The crops of the latter were filled with turnip (not tops), various field seeds, and a little grain. One contained thirty-seven sprouting beans, weighing nearly 1½oz., besides some grain. We replanted the beans, which in due time grew to maturity.

Previous to their appearance as above described, the only Stock-dove observed here was one shot by my brother on September 25, 1878; and in 1881 a pair built a nest in the ivy on the side of the house, and laid one egg on April 18.

This nest was subsequently deserted. I should, however, add that on the moors I have known of the Stock-doves nesting regularly for a much longer period than the dates above mentioned. They arrive there in March, and nest in crags and under boulders, out on the open moor.

During the present winter we have again had the Stock-doves; but it has been an unusually bad season for pigeons generally, and the number obtained is insufficient to form a criterion as to whether we can regard the influx of Stock-doves as permanent. Of the Wood-pigeons, the only arrival of any note occurred on December 28, when very great numbers appeared. It was a squally, boisterous day, wind west or south-west, and all the morning, commencing at 8 A.M., the pigeons kept pouring in in packs of a score or two up to hundreds. The packs appeared at frequent intervals, all coming from the eastward, apparently direct off the sea (three miles distant). They were all clean, light-coloured birds, making the woods look quite white where they perched; but, though I carefully abstained from molesting them, they all passed on westward at night.

BIRD-LIFE OF THE BORDERS

WILDFOWL OF THE NORTH-EAST COAST:

THEIR HAUNTS AND HABITS.

WILDFOWLING WITH THE STANCHION-GUN.

That the enthusiasm even of Colonel Hawker, in his masterly exposition of the art of wildfowling afloat, has failed to popularize this branch of sport, appears to show that the somewhat arduous conditions and that inherent element of uncertainty (which to its disciples are the very cream of the whole thing) will ever preclude more than a limited section of the shooting world from enjoying a practical acquaintance with this pursuit. To follow it up by day and by night in winter certainly entails some personal inconveniences, and perhaps even hardships. Probably, too, it demands a combination of mental and muscular qualities which are not often conjoined. Keenness, patience, and untiring perseverance go without saying; but beyond these at least a degree of hardihood and physical strength are essential, as well as a cast of mind able to endure with equanimity the inevitable failures, and those disappointments which so frequently occur from causes beyond the fowler's control.

Distaste, or lack of the requisite qualifications, is, however, no valid reason for attempting to cast aspersions on the sport. It has always appeared to me a matter for regret that writers like St. John, otherwise so well acquainted with his subject, should go out of their way to throw a slur upon a branch of sport in the practice of which they were wholly unversed. Since his day, almost every casual writer on sport has thought it necessary to follow on the same lines, and inveigh vaguely against the pursuit as one "unworthy of the name of sport" . . . . "only an occupation for fishermen," and such like. The general drift of the ideas of these writers appears to underlie an erroneous assumption that (1) it must be a simple business, given a large gun and "dense masses" of fowl, to kill quite an indefinite number; and (2) that, as firing, say, a pound of duck-shot into the "thick" of a number of birds necessarily wounds as well as kills, the proceeding is therefore cruel and inconsistent with proper sportsmanlike instinct.

Now the question of cruelty in sport is not one that I intend to go into here. Cruelty is part of the set order of things as ordained by Nature. Every form of life is preyed upon by some other form ; not a creature but is the destined food of some other; throughout the animal world this rule holds universally, and man, though the consummation of that world, is after all but a predatory and carnivorous animal. There must necessarily be some cruelty in sport, and in many other of the affairs of life; but that the element of cruelty enters into the pursuit of punt-gunning to a greater extent than is the case with many other forms of sport I must unhesitatingly deny. Without being suspected of wishing to animadvert on any branch of sport, I am certain that the loss of wounded game--say in Grouse-driving, covert-shooting, or ferreting—is quite as great, proportionately, as in coast fowling with a stanchion-gun. Consider the question for a moment. During a long day or night's work, the punter gets on an average, say, two shots—four is the utmost I ever obtained myself. After each shot he knows full well there will be no more fowl in his immediate neighbourhood; consequently all attention is concentrated on securing the "cripples." As long as a wounded or disabled fowl remains ungathered, every exertion, ocular and muscular, is strained to get it on board. It nevertheless undoubtedly happens that, despite the utmost efforts, a certain proportion of wounded fowl do escape at first, especially by night. These are mostly but slightly-struck birds, which perhaps showed little or no signs of having been hit. There is, however, another element which tends greatly to minimize the sufferings of any escaped "pensioners." It must be remembered that the moment a bird is wounded, it separates from the rest, and becomes a comparatively easy prey to the first gunner who chances to fall in with it. Hence, in any harbour or estuary well frequented by fowl and fowlers, there turns up during favourable seasons, and especially in severe weather, an abundant incursion of shore-shooters and small-boat gunners, all keenly on the alert to secure any spoils from the punt-guns that may fall to their share. The shore-men are mostly farm-hands, temporarily thrown out of work by the frost and snow, to whom a pair or two of ducks, or a fat goose, form no small addition to the *res angustæ domi.* As to the boat-sailers, one perhaps entertains less charitable feelings. The waters are infested with them—every idle loafer who can beg, borrow, or steal a boat and gun, or even a boat alone, with which they are nothing loath to run in right before the puntsman's eyes and pick up his "droppers." Notwithstanding, however, the nefarious tendencies of this latter class, these small-gunners do at least minimize the chance of any badly-wounded bird being left long in pain; and, on the whole, I am inclined to question whether any considerable number of such fowl ever really escape. Certainly no more proportionately than inevitably do so in field sports, and I have witnessed them all.

Even those few which at first may succeed in escaping, are unlikely long to evade the scrutinizing eye of the Great Black-back and Glaucous Gulls, ever keenly on the look-out for the flotsam and jetsam of the waves, and whose ravenous maws speedily put any sufferers out of pain.

I now come to the second part of my subject, "the pound of shot and unsuspecting masses" theory. Much of what has been written in this connection is closely akin to those foolish diatribes so often penned against modern covert-shooting by those who never in their lives negotiated a "rocketer." Both are pure and simple appeals to popular ignorance. It is the blind leading the blind; but to the *cognoscenti* alone is the full depth of error and prejudice to which they plunge clearly apparent. Place one of these self-confident critics below the wind at a covert-end, or send him afloat, single-handed, in a gunning-punt: half a day's practical experience will then silence for ever both the cant about "tame, stupid, hand-reared, domestic fowls" in the one case, and the flow of cheap invective in the other. As regards the latter, the critic will find that "dense masses" of wild fowl are not so simple and unsuspicious as he had ignorantly assumed; that, in point of fact, he is wholly impotent to approach them; moreover, nineteen shots out of twenty are not at *dense masses*, but at very small companies—often under a dozen. He will find that a gunning-punt is not so easily held in hand as may have appeared to be the case; and that a stanchion-gun is not only a somewhat unwieldy weapon to manage, but that unless it is managed (and managed, too, with a degree of skill which only a fairly long apprenticeship will secure), it proves in his hands of no more effect than a barrel-organ.

Some of the allusions to the use of a stanchion-gun as "murder made easy," rather remind one of the cartoons which so delighted the Parisian mob in the summer of 1870. A French soldier was represented turning the handle of a mitrailleuse, the background strewn with Teutonic dead. Jean had just received orders to cease firing, and turning round, innocently asks, "What! are there no more left to kill?" Fowl too, like Prussians, are easily enough killed *on paper*, with a big gun; but in actual practice, our contemners might find these weapons as deceptive as the mitrailleuses themselves afterwards proved at Wöerth and at Gravelotte.

The successful handling of a stanchion-gun, so far from being the simple matter inferred, is a profession capable of considerable development in experienced hands; and to use these guns effectively is at least as difficult as any ordinary shooting from the shoulder. By effective use I mean the placing of the charge, at precisely the right moment, at the point of greatest advantage. A "duffer" may manage now and then to kill a pair or two of fowl in a haphazard sort of fashion; but the difference between him and the skilled gunner is, that the latter would probably in the same circumstances have bagged a dozen.

One factor in perpetuating the erroneous impressions alluded to is the difficulty of putting down in black and white the salient features of the punt-gunner's craft, or of diagnosing the points which constitute his skill. They can hardly, indeed, be reduced to paper, any more than dexterity in this or in any other sport can be acquired except by practice and experience. After jotting down ideas for years and filling page after page of note-books on this subject, I now strike the pen through the whole, and will confine myself to a few remarks on the initial difficulties which an aspiring punt-gunner may expect to encounter.

When "setting up" to fowl, the sportsman, lying flat on his chest on the bottom boards of his punt and with his eye only raised a few inches above sea-level, has the whole watery horizon for several miles compressed into a (vertical) space of, say, 1ft. Between his eye and the birds are interposed, say, 8in. vertically, equalling perhaps 200 yards horizontally, of rippling grey waves, amidst which the thin grey and interrupted line of fowl appears and disappears twenty times in a minute. The difficulty of correctly judging distance under these conditions, and from so prostrate a point of view, is obvious. There is not a single mark or guide to assist the eye. The 200 yards may appear only 60, or 60 be mistaken for 200, under varying atmospheric conditions; yet correct judgment of distance is one of the first and most essential elements towards securing a really successful shot.

Well, we will suppose our friend has managed to discriminate the distance between his eye and those grey dots bobbing about among the rippling grey waves; and also that he has so far succeeded as to approach to some 60 or 70 yards' range. At this distance a difference of but 2in. in the sighting or elevation of a 6ft. barrel makes a difference of nearly 6ft. in the line of the shot. But, on running his eye along the gun to take aim, he will probably find that there is at least that amount of perpetual motion on the muzzle, owing to the "life" of his craft in the sea. Thus the gun lifts one moment far over the heads of the fowl, the next dips as far below them. Then, while striving to make sure of his level, he perhaps finds the thick clump at which he was aiming has melted away, and the boat's head must be brought round to bear on another clump a point or two away. While executing this manœuvre, the nearer stragglers begin to raise their heads—to crane up their necks. They are going to fly. The gunner knows it is now a matter of a second or two; perhaps gets flurried; and the upshot of the affair is often that he fires at random, with long odds on a clean miss, or that the birds fly up, when, if Ducks (which spring high), his whole charge passes harmlessly beneath them.

The stanchion-gun, in short, does not lend itself kindly to accurate sighting when afloat—that is, as sights are ordinarily taken. Accuracy of aim is required even to secure a half-grown young rabbit placidly cropping the grass on a summer evening when close at hand. How much more necessary it obviously must be effectively to reach wildfowl at perhaps nearly a hundred yards' distance on the sea! Yet a skilful fowler will, in a majority of cases, succeed in placing his charge at or near the point of greatest advantage, though the means by which he does so are hardly reducible to paper. They undoubtedly comprise a combination of technical experience, judgment, and decision in action, together with coolness and the absence of that flurry which is often provoked by the near propinquity of large numbers of fowl.

But enough of the argumentative and critical; let us glance at the sport from another point of view. To a lover of Nature, or to any one who delights in the observation of wild-life, a gunning-punt affords unrivalled—indeed unique—opportunities for studying, at moderately close quarters, some of the wildest creatures in existence. So intensely wild, indeed, and so intolerant of the human presence, are many kinds of our winter wildfowl, that, except by this means, it is a simple impossibility to form even a remote acquaintance with them, or to observe their life-habits and interesting idiosyncrasies. While, as to obtaining them, despite all the foolish things that have been written, the use of the stanchion-gun is an absolute *sine quâ non*. A wildfowling trip to the coast with a shoulder-gun is usually a mere farce, and results, as Colonel Hawker well remarked, "in a £10 bill, and perhaps bagging a couple of sea-gulls."

Probably no other class of our British fauna is less precisely understood, or presents more promising material for investigation, than the numerous family of wildfowl. To a mind appreciative of such subjects, therefore, the winter months on the coast possess abundant attraction. There, on the open salt water, in a favourable place and season, very great quantities, and no small variety, of wildfowl can generally be seen. I write *seen* advisedly, for to few is it given to handle them comprehensively; and, as a matter of mere sport, the results of casual coast-shooting are seldom considered commensurate with its difficulties and hardships—as a temporary abandonment of luxury is called. Even to a fully equipped puntsman there is seldom a certainty of success. When one starts for a day's shooting at home, or sets out for a week's campaign on the moors, the probable bag can usually be foretold with tolerable accuracy. There may be a few brace more, or a few brace less; but on the salt water no such forecast is possible. So restless and shifty are wildfowl—here to-day, gone to-morrow—so bleak, exposed, and shelterless their chosen haunts, and so many and fickle the ever-changing conditions of time and tide, wind and wave, on which the fowler's fortunes are wholly dependent, that, however much he may deserve, he cannot command success. Let him always start brimful of confidence, but ever prepared to face total failure without a murmur.

During fine, mild weather in open seasons, when the fowl have no difficulty in obtaining both food and rest, they often become wholly inaccessible. The fullest practical knowledge, abetted by the most approved appliances* of boat, gun, and gear yet devised by man, are not seldom under these conditions of no avail, and prove unequal to the task of circumventing a single fowl, though thousands may be in sight. These climatic conditions may, and do, prevail for weeks, months, or even a whole season, when hardly a dozen fair shots will be obtained all the winter. I mention this fact, as some pretentious authors have undertaken to tell us how fowl may be got under all or any conditions.

Sooner or later, however, the chance will come. It is in the hardest winters, when for weeks at a time the landscape lies buried under a deep mantle of snow—in severe and long-

* This remark may perhaps not be applicable to the latest developments in huge double swivel-guns. I only referred to the punt-guns ordinarily used, not having then seen the monsters alluded to—the costliness of which will effectually prevent their coming into general use.

protracted frosts, when the salt water and the oozes freeze between tides—at such times it is that the wary fowl can at last be got. Then very few people care to brave it on the coast or to undergo what they would probably describe as "the hardship of being out for hours in a piercing cold—turning night into day, and day into night—lying cramped up in a damp raft as narrow as a coffin, constantly wet, and often in actual danger," &c. To the true wildfowler, enamoured of his craft, such seasons are the acme of enjoyment. To him the difficulty of getting these birds, their beauty of form and cry, the great variety of their plumage at different stages—ay, and the very hardships themselves—are ever superlatively attractive. Personally, the writer may fairly say he seldom feels so happy as when on board his trim and smart little craft, gliding smoothly and rapidly over the icy waters which lipper and play over her sharp bows, and dance along the white and rounded decks. The prospect around may be dreary—heaven and earth blending in long vistas of flat oozy "slakes," the monotony only relieved by barren, bent-grown sand-links. But there is a charm in the solitude and in the feeling that one is monarch of all one surveys, and has inherited an innate dominion over the fowls of the air (a sentiment, by the way, which is often rudely dispelled); yet the presence of the beautiful creatures on wing or water around can transform these dismal mud-flats into one of the most delightful resorts. The world, in fact, is very much what each of us makes it for himself.

WILDFOWL OF THE NORTH-EAST COAST:

THEIR HAUNTS AND HABITS.

IN physical conformation the north-east coast of England is not well adapted to the requirements of wildfowl. Geographically, no doubt, it forms the objective point of the trans-oceanic journeys of a large proportion of those migratory hosts from northern Europe and Asia which, every autumn, direct their flight upon our islands. Great numbers of these certainly "make the land" within the limits so defined; but there is little attraction to induce them to remain here. They are aware, or soon discover, the deficiency of suitable resorts congenial to their tastes; and, in consequence, after brief periods of rest, move on to localities offering a greater measure of their desiderata. These chance casual visits, or "through transits," of wildfowl are of but little value to the fowler, occurring so irregularly, and without calculable or presumptive fixity of date. There is no "North-Sea Bradshaw" available to disclose their probable arrivals or departures. Hence the pursuit of wildfowling in the north-east is limited to a few enthusiasts, and is carried on more as a matter of local convenience than otherwise—it may be said to be pursued rather in spite of disadvantages than by reason of any special facilities which this coast affords.

The north-eastern seaboard is too straight and exposed, and its configuration is wanting in those irregularities of outline which denote sheltered bays and land-locked waters, the abundance of which on a map—say, of western Ireland —give that coast so attractive an appearance to the eye of a wildfowler. The coast line from the Humber to the Forth is a series of almost straight lines, and a great proportion of this extent is occupied by lofty cliffs, rising sheer from the sea. Even where the shores are low and flat, the lines are so straight as to leave no extent of foreshore—that is, the space between high and low water-mark is merely a long strip of sand, shingle, or rock, only a few hundred yards in width at the utmost. Neither of such situations are at all congenial to the requirements of wildfowl, properly so called. The sandy stretches attract a few small waders, and the cliffs afford suitable homes for the Gulls, Guillemots, and other rock-birds. Flamborough Head, the Farne Islands, and the beetling grey precipices of St. Abb's Head, in Berwickshire, are notable breeding resorts of those birds; but such iron-bound coasts are the last places in the world for true wildfowl.

Then, too, the encroachment by man on the foreshores has seriously interfered with the few localities which, in former days, attracted large numbers of fowl to this coast. The development of the northern coal and iron trades has transformed what fifty years ago were desolate tidal wastes into busy scenes of human industry—their once-deserted shores now flanked by towns, docks, and factories, with their concomitants of smoky chimneys and the other paraphernalia of "civilization." From such places the altered conditions and the incessant turmoil of revolving paddles and propellers have effectually banished the fowl—never to return. Such a spot is Jarrow slake, on the Tyne, and the Teesmouth is rapidly following suit.

The places which are the most favoured haunts of wildfowl are precisely those which are least frequented and least congenial to man—the most remote and lonely expanses of tidal ooze. Such conditions usually only prevail either at the estuaries of large rivers, or on those low-lying parts of the coast where land and sea are engaged along the boundary line in one ceaseless perennial struggle for dominion—their battle-ground a vast level stretch of sand, mud, and ooze, which *nec tellus est, nec mare*. In such a spot, at low tide, the eye roams over an almost illimitable expanse of flat, featureless foreshore; miles away in the far distance, across the mud-flats, and across the broad yellow sand-bar beyond, the white line of the breaking surf is just distinguishable against the grey background of the open sea. In a few hours time—at high tide—this whole expanse will be one great sheet of blue salt water, reaching right up to the hedgerows of the stubbles and pasture fields, and to a passer-by not distinguishable from the sea itself. To all appearance it might be forty fathoms deep; yet, as a matter of fact, there are thousands of acres over which the maximum depth—except in a few tide channels—never exceeds from three to six feet, and of this depth one-half or more is occupied by the long waving fronds of the sea-grass growing from the bottom.

Such a place is the *beau ideal* of a wildfowl resort; those tiny white dots stretching far along the shore are a couple of thousand Geese. They are in five feet of water, but graze easily on the long shoots of the sea-grass beneath them. To such a resort as described wildfowl still come every winter in undiminished numbers (greater or less according to the season), and will continue to do so as long as such places continue to exist, despite all that man can do. His persecution, with all the artifices which his ingenuity can devise, has certainly not the slightest effect on their numbers, though it unquestionably modifies many of their habits, as I propose hereafter to show. It should be remembered that twice every twenty-four hours the fowl have secured to them by the ebb of the tide periods of several hours' absolute immunity from molestation; for then they can feed or rest, in absolute peace and security, right out in the centre of miles of mud-flats far too solid to admit the approach of a punt, and yet too soft and "rotten" to bear the weight of a man.

The staple fowl pursued by punt-gunners on the N.E. coast (as in most British waters) in winter, are Brent Geese by day, Mallard and Wigeon by night, Teal being seldom met with on salt water after their autumnal passage in September and October. Among the minor objects of pursuit are the diving-ducks, chiefly Scaup and Golden-eye, as well as the larger class of wading birds—"hen-footed-fowl" I have heard them called—both of which are mostly obtained by day. There is, however, some variation in the sport, according to locality. Thus, while Wigeon frequent every considerable estuary which is undisturbed and otherwise suitable to their habits, the Geese are far more captious, frequenting one harbour year after year in very great numbers, while another, perhaps only a few miles away, and apparently similar in its natural features, is never entered by them.

In order to ascertain what fowl frequent any particular harbour or bay, it is only necessary to watch their "morning flight" on two or three occasions. The whole stock of fowl may then be observed, from a favourable post, in the course of a couple of hours. The time thus to "take stock" is at an hour about the break of day—sooner or later, according to the tide. The tide, in fact, on the coast supplants to a great extent the ordinary chronological measurements in vogue elsewhere—as, somewhere about the middle watches of the night, a burly fisherman ruthlessly obtrudes on the deliciously unconscious sleep of the weary fowler, and one helplessly asks him what hour it may be, the reply is, "It's quarter-flood, sir, and there's no time to be lost!"

As the first streak of dawn becomes discernible in the eastern skies—or rather a little before that period—there commences a general movement of wildfowl, and from a favourable position (usually near the mouth of the seaward channel) the whole local stock of fowl may be observed in the course of an hour or two's watching—the night-feeding birds speeding outwards to the open sea, and those of day hurrying in, hungry, to their feeding grounds within the harbour. Lying concealed among the weed-covered rocks of the outermost promontory, the gunner enjoys a moving panorama of bird-life which amply repays the trouble of turning out an hour or two earlier than usual. The nearer the water's edge he lies, the better his chance of a shot; and he can shift his position a few yards backwards at intervals, as the tide creeps up to his sea-boots.

At first it is pitch dark, the rude features of the coast scenery but dimly discernible, and only the wild cry of some seafowl heard blended with the roar of the breakers outside.

First to move are the Mallards, then the Wigeon ; both of these, in winter, go out to sea before a symptom of daylight has appeared. They are only recognizable by their well-known notes (if uttered), or by the resonant swish, swish of their strong pinions, distinctly audible far up in the dark skies. Perhaps the stately lines of the Mallards may be discerned for a moment against some cloud-bank—Wigeon never form line, but hurry out in a confused mass. Next to these come the Mergansers, the first of the "inward-bound" from the sea. They come singly, or in twos and threes, flying very close, as though linked together, and at a tremendous speed. Then the darkness resounds with the noisy vibrations of a thousand wings, as a dense, shapeless mass of Godwits or Knots rush past from inside, or a string of Oyster-catchers pass overhead—all these waders being driven out as the sand-banks disappear under the flowing tide. The latter—Sea Pyots as they are called—bear a strong resemblance to Duck as they file out in line, and in the uncertain light many a one has lost his life, owing to this unfortunate similitude to his superiors. The waders are not, of course, bound for the sea, but for some extensive salt-marsh or sand-flats they wot of along shore, where they can rest in security during high water.

As the light gradually strengthens towards the dawn, great spectral forms loom silently overhead; these are the big gulls diligently searching the waters for their breakfast, and the boisterous laughter of the small Black-headed Gull resounds from the tideway beyond the bar. Early one January morning a huge Glaucous Gull settled down on the water close at hand, carried off one dead Godwit, and deliberately pulled another to bits—two out of half a dozen that had just fallen to Mr. Crawhall's gun.

Next a Grebe may perhaps come spinning along. Close to the water he flies, and, considering the shortness of his wings, at an amazing speed. Then a few Golden-eyes, usually singly, and always very high, pass inwards. Meanwhile the Geese are on the move; and, in the dim light seawards, one descries, far away over the dark waters, what might be the edge of a little cloud, or the smoke of a distant steamer, as they go through their matutinal evolutions preparatory to "coming inside." For some ten minutes these evolutions continue, and in the increasing light the forms of their dense columns become gradually discernible, gyrating rapidly to and fro beyond the line of breakers to seaward. Presently up they come, always high in air, unless half a gale blows right in their teeth, and pass up the channel, clanging down, as it were, a loud defiance to man to do his worst. As daylight becomes fully established, there appear the weird-looking Divers (*Colymbi*), usually the last, or perhaps an unwieldy Cormorant brings up the rear; and now the rim of the sun appears above the eastern horizon, and one lingers a few minutes longer while the eye revels in the gorgeous hues and the lovely effects of a sunrise over the sea.

Such, in rough outline, is the "morning flight," as it may be observed any winter's dawn at a well-frequented resort. The fowl pass in and out pretty regularly in something like the order named, and to a lover of bird-life the whole scene is a delightful and interesting one. In addition to the species named, various others may be observed, according to locality. Thus, for example, the Grey Geese pass certain points as regularly as dawn and dusk come round. The particular harbour the writer has in his mind's eye in describing the above, does not happen to lie in their line of flight, though they pass regularly over a point only a few miles distant.

Apart from the charm of observing these wild creatures, there is but little other reward, for it is seldom that any of the more desirable fowl pass over within gunshot; on fine, calm mornings especially they fly very high—up in the clouds. During rough, boisterous weather, when the force of the wind is dead against the fowl, their flight is lower, and on such mornings a pair or two *may* be secured before breakfast. As it is precisely on such occasions that one cannot go afloat in a gunning-punt, the morning flight then affords an interesting, and sometimes exciting, means of relieving, for an hour or two, the tedium of what would otherwise be a blank day.

A good eye and some judgment are required to distinguish the different species of fowl as they speed overhead. The local gunners of the coast have often remarkably good knowledge of these points; the casual amateur little or none. I recollect, a few years ago, one of these latter displaying with much pride what he called "a Wigeon" he had shot. When the palpable fact was pointed out that the bird was not a Wigeon, but a common Mallard drake, he remarked, "Well, that's what we call a Wigeon *in the south*, at any rate!" Next morning our friend jauntily strolled in to breakfast with a pair of Brent Geese slung over the barrels of his 12-bore, and ostensibly slain therewith. Such success —the morning being dead calm—was remarkable; but during the day I chanced to discover that it was traceable to the propinquity of a professional fowler's cottage among the sand-links he affected!

There is some art in using a large gun and killing long shots from a prostrate position at fowl passing fast and high overhead, especially in the half-light. All wildfowl fly fast, some at a tremendous speed; and even those, such as Geese, which appear to go slowly, are travelling far faster than they seem. Many a first-rate game shot, who, when erect on his legs, may be tolerably sure of his right and left under all reasonable conditions, is wofully disappointed in his first attempts to use a big gun. The new conditions and the constrained position, lying prone among sharp and angular rocks, are apt to disconcert his wonted skill; and he has perhaps to admit himself far outstripped at the business by the long single barrel of a fisherman-gunner, who probably could not hit a partridge to save his life.

WILDFOWL OF THE NORTH-EAST COAST:

THEIR HAUNTS AND HABITS—(*continued*).

HAVING obtained from the "morning flight" a tolerably accurate idea both of the numbers and the variety of wildfowl in the neighbourhood, and of their distribution for the day, we will launch the gunning-punt and follow that section of the fowl which have passed *inwards* (*i.e.*, up the harbour), leaving those which have gone out to sea for another day.

As the flowing tide covers the flats and the punt glides over what had just before been a vast plain of slimy ooze, sprinkled all over with a greenish-looking garbage, one is surprised to see, beneath the craft, a luxuriant mass of foliage. The mud has completely disappeared beneath a dense growth of long green grass waving to and fro in the tide-currents like a rich crop of clover-seeds on a windy day in June. This sea-grass, so graceful when submerged, so uninviting when lying high and dry on the ooze at low tide, is the *Zostera marina*; it is the first essential of wildfowl. What heather is to Grouse and the stubbles to Partridge, such is the *Zostera* to our sea-game. Geese and most of the surface-feeding ducks live almost exclusively upon it while on our coasts, and to the broad expanses of mud and ooze where it grows abundantly, they will constantly resort, despite all the artifices of the fowler. Here, roughly speaking, the Geese feed by day and the Ducks by night, and will continue to do so as long as such places continue to exist.

The extreme luxuriance of this sea-grass is such that over thousands of acres its densely thick fronds, each measuring four or five feet in length, completely cover the whole surface of the ooze as closely as the grass in a meadow. The depth

BRENT GEESE.

To face page 162.

and rich quality of the mud itself are favourable to this exuberant fertility both of the *Zostera* and of many algæ and other marine plants which grow on its surface. Among these is the marsh samphire (*Salicornia herbacea*), which alone stands upright, not unlike the "mare's tail," but does not appear to be relished by wildfowl. The profusion and variety of marine vegetation which flourishes in such situations is, indeed, as great as that which clothes the inland fields and fells, and it is this which attracts the wildfowl to our coasts. Yet how strange is the almost universal error that wildfowl live on fish. "What can you do with these ducks and geese; surely they must be very fishy?" are the almost invariable questions asked, often by people who should know better. The food of the game-ducks and geese is quite as exclusively vegetable as is that of the game-birds themselves.

When cruising about in these wide flat "salt slakes," one soon observes that they consist of two different materials, each possessing very distinct characteristics. The two materials are MUD and SAND. On the former alone grows the *Zostera*, and to it, therefore, resort the flat-billed fowl —ducks and geese. But the sand-flats and the sandy-bottomed channels have each a character and a fauna of their own. The bird population of the sands are chiefly waders—Curlews, Godwits, Grey Plovers, and birds of that ilk, together with a few Sheld Ducks; while the deep-water channels, or "guts," are the resort of divers of all denominations, namely, Scaup and Golden-eyes, Mergansers, Cormorants, Grebes, and the Colymbi.

The sand-flats, which on some parts of the coast are of immense extent, are not favourable grounds for punting operations, even though occupied by thousands of birds. Their surface is too level. A fall of, perhaps, half a foot to the mile renders it demonstrably impossible to float a punt (drawing, say, three inches) within many hundred yards of birds sitting on the "dry." The mud, one would think, is flat enough; but on it there *are* slight banks and hollows, and shallow winding creeks. Sand is a far less coherent substance, and the powerful sweep of the tide removes the very slightest irregularity of surface, and reduces the whole wide area to one practically horizontal plane as smooth as a polished mahogany table. When cruising along the edge of such a place, what looks like a solid wall of birds may be observed ahead, from which there resounds a tumultuous babble of harsh voices. These are all Godwits, standing in very shallow water. Long ere the punt can float within shot, the solid wall is seen to be melting away by driblets—now a dozen, then fifty, or a hundred birds depart in detachments as the tide creeps up, and before a fair range can be attained, hardly a pair can be observed together.

All these larger waders are feeding chiefly on the sand-worms, whose numbers in such places are simply legion. Hardly has the tide ebbed off the sand than its smooth surface is dotted and spotted all over with their myriad little castings—as many as fifty to a hundred in the square yard. Out beyond these flats to seaward, and separating them from the open sea, there lies, in most harbours, the sand-bar, a region of a different character, great part of which is never covered by the tide—and of which, more anon.

Thus, wildfowl resorts may be roughly divided into three main regions, each of widely different features—(1) The mud-flats, or "slakes," as they are called for distinction; (2) the tidal sand-flats; and (3) the non-tidal sand-bar.

Hard weather, as already mentioned, is unquestionably the time of all others to see wildfowl in their most profuse abundance; it is, moreover, the time to get them. Severe frost, it is well known, has a deadening, or soporific, effect on all (or nearly all) fowl, rendering them less alert to threatening dangers, and it also produces certain changes in their normal habits, engendered by the altered climatic conditions and by the difficulty in obtaining food. Mallard and Wigeon, for example, will, under these conditions, forsake their regular nocturnal habits and resort by day in large numbers to the oozes and feeding grounds of the estuaries. Here they are, then, comparatively easy of approach, especially the Mallard, on which severe cold produces exceptional effect. Small "paddlings" of six or eight to a dozen may be approached in a gunning-punt within forty yards, sitting apparently fast asleep among the ice, and with their bills

"ON THE FLATS." SEPTEMBER.

To face page 164.

snugly tucked away among their back feathers. Many of these modifications of the habits of wildfowl during severe weather, and especially the precise causes which tend to produce them, are a most interesting subject for investigation.

The sand-flats, just described, afford one main factor in bringing about the altered physical conditions which then prevail. Over these vast dead-level expanses of wet sand there is spread a thin covering—a mere film—of salt water, either left by the tide or blown up by the wind. In severe frost this film freezes into a sheet of ice, so thin and elastic (or pliant) as to rise with the succeeding tide, unbroken; with every ebb its thickness increases. Then, perhaps, comes a snow-storm, and the surface of the ice is covered three or four inches thick with snow. This rapidly freezes to the ice below, and, in fact, forms a compact and homogeneous mass with it. Thus in the course of a severe three days' frost there is formed on the sand-flats what can only be described as one vast ice-field, of perhaps many hundreds of acres in extent, and five or six inches in thickness. The weight of the ice gradually causes the solid field to split and crack up on the flood-tide, the chief breakages occurring along the outer side, where the water is deepest and the influence of the tideway most felt. The whole ice-field also tends to drift outwards on the ebb, and, as long as the frost holds, the process of congelation and the creation of new ice continues, and goes on afresh. Thus very great quantities of ice are carried off the flats by every ebb, while the places left vacant are being rapidly reoccupied by a further generation of glacial supplies.

Even in a frost of but a few days' duration the quantity of ice thus daily set afloat by the tide is very considerable. But when, as happens in such extremity of cold as we experienced in the winters of 1878-9, and again in January, 1881, the frost continues unbroken for weeks at a time, the phenomena created thereby are indeed almost incredible, save to those who have witnessed them. The masses of detached ice, split up by their own weight into fragments of all sizes and shapes, and carried here and there by the currents, drive helter-skelter in the tideway, and along the lee shores are thrown up into ridges and rugged piles, extending for miles along the shore. Outside this glacial barrier of stranded blocks, the floating floes, carried along by the strong tide-currents, grind and crash against each other, piling up cake upon cake till they become miniature icebergs, and form a spectacle such as few have seen outside the Arctic regions.*

The effect of such metamorphoses upon the fowl are obvious; great part of their feeding grounds are rendered inaccessible to them. At low water, or during neap tides, many hundreds of acres of the mud are buried under the stranded ice-floes, and as the tide rises the whole area is occupied by these blocks, rushing, driving, and careering forward on the current. So great is the turmoil that it is in the highest degree unsafe to venture among the moving ice in a gunning-punt; and, of course, for the fowl it is even more impossible safely to visit their accustomed feeding haunts. Especially must this be the case by night, when the drift ice would be invisible; hence many of the nocturnal fowl are at such times to be seen about the oozes by day.

It is at such times as these, when ice and snow cover both land and sea, and the angry leaden sky is spangled with driving feathery flakes—at such times one may look for the appearance of the Hooper, monarch of the flood, and several of these noble wildfowl sometimes pay the penalty of their lives ere they learn the wisdom of giving a wide berth to that low, white, unsuspicious-looking craft which so closely resembles the blocks of ice drifting along on the tide. In the mild winter of 1882, about Christmas, several of the smaller species (Bewick's Swan) appeared on this coast, and, being very incautious, were, I believe, all killed.

* An apparently similar result is produced by heavy snowfalls, even when unaccompanied by hard frost. The snow which accumulates during low water on the flats becomes solidified by the rising tide, and floats away in large blocks closely resembling ice. These, however, are comparatively soft and "rotten," and can be penetrated by a punt without danger, and with but little trouble. We had plenty of this half-frozen drift snow during the early part of the month of January, 1887

In such seasons, the number of Brent Geese on this coast, as mentioned elsewhere, is truly amazing. In the winters of 1878-9, and 1880-1, it was no uncommon occurrence to see from ten to twelve thousand in a single harbour; and during March, 1886, even these great numbers were largely exceeded. The soporific effect of the frost on Mallard and Wigeon has already been alluded to. Sheld Ducks (usually rather wary fowl) now become quite tame; indeed, they and all the shell-feeding birds suffer severely in protracted frosts. The vegetable feeders can make shift for a considerable time on the drift weed, which every tide is carried off the flats by the ebb; but the food of the others is absolutely sealed against them, and they suffer proportionately. Even the Curlews, usually quite impracticable, now yield to the extremities of the weather, and may easily be approached, if thought worth shooting in their emaciated and half-starved condition. Scaup-ducks, always tame, will now admit of approach even on foot; and the lesser waders hardly take the trouble to get out of one's way. The effect of long and severe frost, in short, is to tame and subdue all wildfowl, and render them accessible to a punt, though Mergansers and Golden-eyes are always the least affected.

During the continuance of the frost, wildfowl enjoy one great safeguard, in the *ice itself*, from the ascendency which would otherwise accrue at such times to the gunning-punt. As already mentioned, it is most imprudent to adventure these frail craft among the drifting floes, where they run a serious risk of being stove, to say nothing of the impossibility of holding a direct course, or of working a big gun under such conditions. Moreover, during neap tides the accumulations of stranded ice along the "full-sea mark" render large areas of the flats wholly inaccessible to craft of any description. The very best chances to score occur, therefore, not so much during the frost itself as on the first break-up of the ice. The fowl are then so intent on getting a "square meal," and so determined to make up for the hardships and short commons of the "glacial epoch," that excellent opportunities may be secured by those who are lucky enough to be on the spot at exactly the right moment.

Within thirty-six hours of the break-up of the frost, every vestige of ice has disappeared, carried off to sea by the tide, and normal conditions are at once restored.

Incidentally, I may remark that the effect of very low temperatures on the human body when exposed for many hours together in a punt at sea, is relatively much less severe than one would expect. Of course, one must be suitably clad. Abundance of warm woollen clothing goes without saying; and no part of one's flesh, except what is actually necessary, must be exposed to the bite of the frost. The only limit as regards nether garments is the capacity of the sea boots; and, as to upper gear, the ability to handle the cripple-stopper.

Whether the effect is caused by the relatively higher temperature of the salt water, or by the extreme dryness of the atmosphere owing to the entire absorption of all moisture by the frost, or otherwise, I can state from experience that one suffers a great deal less from exposure to cold in the lowest known temperature (as in January, 1881, when for several days the thermometer stood from 2° to 7° below zero, and the salt water at once froze solid on the setting pole, and even on the rounded decks of the punt) than is the case in the raw, chilling, marrow-piercing winds of a "mild winter."

"ARCTIC NORTHUMBERLAND.

To face page 168.

THE GAME-DUCKS.

UNDER the head of "Game-ducks" there are included, from a wildfowler's point of view, those members of the duck tribe which are met with on the sheltered waters of land-locked estuaries—in other words, on waters where a gunning-punt can be safely navigated. These are chiefly the surface-feeding ducks, such as Mallard, Wigeon, &c., which feed by night; but a certain section of the diving ducks, such as Scaup, Golden-eye, &c., also frequent these same situations by day, and must therefore be regarded as forming a sub-division of the Game-ducks. The remaining section of diving ducks, being confined exclusively to the open sea, are seldom or never met with in waters where a gunning-punt can venture. These are classified, in wildfowling parlance, as sea-ducks, and to them I will refer in a subsequent chapter.

By far the most important of the ducks included in the first-named category are the Mallard and Wigeon. From September till March they are both the most numerous, the most valuable, and the most sporting fowl on the coast-gunner's game list. To him these two species are just what the Partridge and Pheasant are to the inland sportsman, and, correspondingly, the Brent Geese on the coast take the place of the Grouse on the heather. The habits of the Mallard when on the coast vary to some extent from those of all their congeners; it will, in fact, be found in every case, when closely examined, that each species possesses individual characteristics peculiar to itself. The different conditions of natural disposition, food requirements, and general physical economies of every separate species vary so infinitely, that a close study of these conditions and of their effects on the respective birds are both extremely instructive and an interesting complement to the pursuit of wildfowling.

By nature the Mallard is essentially and absolutely a night-feeding bird (far more so than the Wigeon); is almost omnivorous in its taste, but with a partiality for fresh water if easily accessible; has a strong inclination to rest by day, but careless as to whether it rests ashore or afloat. Well aware of the danger of remaining inside harbour by day, the Mallards, with the Wigeon, take flight from their feeding grounds, as a rule, before a sign of daylight has appeared. Their most favoured resorts for whiling away the hours of daylight are either (1) on the open sea, opposite their feeding

grounds if smooth, or, otherwise, some sheltered bay or roadstead along the coast, possibly several miles away; or (2) among the tidal channels and shallow backwaters, formed by the tide, in the sand-bars which inclose most large estuaries, or wildfowl resorts, both in this and other countries. Wigeon seldom care for these latter resorts, or to stop short of the open sea.

Of course, if there should happen to be in the neighbourhood of their feeding grounds an inland lake or pool, undisturbed and of sufficient extent, this would be the grand resort of the Mallard (and Wigeon too); but I am now referring exclusively to their habits on the coast.

The sand-bars above mentioned, being one of the characteristic features of wildfowl resorts, deserve a few words of description. In many places these wastes of sand are of immense extent, a considerable portion never being covered, even by the highest spring tides. At low water, mile after mile of flat red-brown sand lies exposed—the result of ages of ceaseless struggle between land and sea. Far away across the level expanse the white line of breakers bounds the horizon to seaward; and one may wander for hours over the yielding spongy surface without an object in sight, except the weather-beaten ribs of some old wreck half swallowed in the shifting sands, or perhaps a big grey seal cautiously basking in the bright October sun, always close to the edge of a deep-water channel. Presently we come across a spot where the smooth, unruffled level is disturbed, and the sand imprinted with the paddlings of many webbed feet. All around lie strewn for half an acre feathers, great and small —many long and strong quills—and other vestiges of a departed multitude. That is where the Grey Geese roosted last night. To-night, if you lie in wait for their arrival, they will perhaps take up their quarters a mile or more away. The flight-gunners bury themselves in the sand at such places as this, on the barest off chance of getting a shot; but, on so vast an area of ground, it is the merest fluke if the Geese should happen to alight within range, and, indeed, the chances of a shot are almost *nil*.

The sand-bars are of course intersected by the main stream of the estuary, which traverses the sand by one or more deep-water channels on its course to the sea outside. On these channels, and along the inner margin of the sand-bar, the strong sweep of the tide, banking up the yielding material, cuts out broad flats on a slightly lower level than that of the sand-bar itself, and these flats, as the tide rises, form shallow backwaters. It is to these backwaters, and to the edges of the main channels, that the Mallard are especially fond of resorting to rest and sleep during the day, and here, during the flowing tide, some excellent shots can now and then be obtained at these fowl, particularly if (as is often the case) the sand-bar lies remote from the quarters of the local punt-gunners. These desert tracts which I have endeavoured to describe are usually neglected by punters, who, as a rule, confine their operations to the mud and ooze, and seldom dream of "poling" perhaps several miles into what appears but a useless waste of sand.

At full tide the area of sand left uncovered is of course greatly reduced, and it is during the last quarter's flood that the best chances are to be secured by the punt-gunner. Then, as the tidal sand-banks one after another disappear, the Ducks which have been dozing the hours away along their edges are set afloat, and, together with those which have been resting in the water of the channels, come driving gently up on the tide, and drift, as it were passively, into the shallow backwaters already mentioned. Here the fowler should be lying ready in his craft, and at such times, barring accidents, is tolerably sure of a fair reward for his patience.

Considering the well-known fact that the Mallard is certainly one of the wildest and most watchful birds in existence, one singular fact has always struck the writer as being among the most inexplicable features in wildfowling, namely, the comparative ease with which these Ducks can often be approached in broad daylight in a gunning-punt. Under such circumstances as I have just described, one sometimes obtains, during a single tide, chances to experiment on a variety of fowl and to compare their relative degrees of wariness. Thus we perhaps first try our skill on a small lot of Grey Geese, resting close to the wash of the sea. No! They won't have it at all! They rise five gunshots away; then some Sheld Ducks fix the range of safety (in their ideas) at three. Even the spread-eagled Cormorants utterly decline negotiation, and as for the waders, they are perhaps the "shiftiest" of all. Incidentally I may remark that the aggregation of these latter birds, which at flood-tide assemble on the sand-bars, is sometimes almost marvellous. The whole interior expanse of mud-flat and ooze up the estuary being then submerged, the wading birds are driven out to the only refuge which remains uncovered —namely, the now comparatively limited area of the sand-bar. At full-sea this resort is thronged, ay! "carpeted," with such multitudes of "hen-footed fowl," as I must decline attempting to describe. To convey an adequate idea of their number would necessitate the employment of inadmissible superlatives, while any estimate would be hopeless. I often wish some of the legislators who premised an Act of Parliament with the preamble that wildfowl are decreasing in the British Islands, could see some of the spectacles I have viewed during the past few months. But attempt to "set up" to these whistling, chattering hordes in a gunning-punt—let your boat be the lowest, the lightest, and the fastest ever launched, and her occupants full masters of their craft—they will utterly fail. The Sea-Pyots, Plovers, and such-like simple birds (if alone) will, no doubt, admit of approach; but as for the rest, the Curlews, Godwits, Knots—well, they know a gunning-punt and its meaning as

MALLARDS ASLEEP—MIDDAY

well as though each of them had a copy of Hawker or Payne-Gallwey in his pocket.

Yet, strange to relate, the Mallards, the finest and most valuable fowl of them all, despite the experience of generations, do not yet seem fully to have learned to recognize the deadly nature of that low white craft. Time after time I have "shoved" up to within sixty, even fifty, yards of their still unconscious flotilla, drifting slowly along on the tide, all inanimate and apparently asleep, hardly a head to be seen. Even after the cruel disappointment of a miss-fire they have not risen at once. Up go their necks, full stretch, at the snap of the cap, and their deep-toned and intensely eloquent "q-u-a-r-k! q-u-a-r-k!" is barely audible, so gently and suspiciously is the alarm note sounded, but they do not rise till one has had *almost* time to replace the cap, but not quite.

There is one imperative point in approaching Mallard, or, indeed, any wildfowl. They must always be "set to" from the leeward; neglect of this precaution assures certain failure. Nevertheless it sometimes happens that fowl are met with in such a situation (as *e.g.*, on the weather side of a shelving bank, or in the leeward recesses of a narrow creek), that there exists no alternative but to attempt to proceed down wind on them. The attempt is sure to fail, but the effect produced on the fowl is interesting to observe. Their sense of smell is obviously keener than their perception of the threatening danger; and as they get our wind when still two or three gunshots distant, the Ducks will be observed to be affected by some suspicion, a suspicion which appears to be vague and to have assumed no very definite form in their minds. The pack begins to scatter, each Duck swimming to and fro uneasily, and when at last they take wing, they do so reluctantly and with slow and horizontal flight, very different from their custom when in presence of ascertained danger. Had one sat upright, fired a gun, or, in short, revealed clearly to them the hated human presence, every Duck would have sprung vertically in the air like so many sky-rockets. Now they fly low, and probably pitch again at no great distance. But that lurking, ill-defined spirit of suspicion continues to smoulder in their breasts, making them restless and shifty, and it is seldom that that pack can again be manœuvred.

So great is the dash with which the Game-ducks (Mallard, Teal, or Wigeon) spring from the sea, throwing themselves at a single impulse a full dozen feet clear of the water, that they cannot well be taken "on the rise" with a set punt-gun. The shot must be taken on the water, or, at latest, just as their wings are opening. Of course, flying shots can be made at Ducks by "tipping" the gun; but these are not so effective, and I refer to the more usual method of firing with the gun laid in position along the fore-deck. To illustrate my meaning, let us compare a shot at, say, Mallard, and one at Geese. The latter fowl, being heavier and less active, rise horizontally from the sea, offering at that moment the most effective shot that can be desired. A gaggle of Geese, when swimming, may be compared to a book lying flat on a table; on rising, it is as though the book was opened in the middle, and half its pages held upright, thus presenting a far more extensive target. All that the gunner need remember on going to Geese is to give his gun a good elevation, and shove ahead full speed till they rise. The Geese themselves, in fact, give the signal when to fire (if within range), and the main difficulty is to attain that distance (no slight difficulty is that, however—one that in a

MALLARD SPRINGING TO SHOT—DAYBREAK.

mild season often proves wholly insuperable). But with Duck the case is different, as their first spring carries them clear of the trajectory of a punt-gun. The choice of the precise moment to fire is, therefore, a fine point requiring both coolness and judgment. There are the proverbial "three courses" open. By firing the moment a long range has been obtained, a couple or two may possibly be secured at each shot; but this is a highly unsatisfactory and most unsportsmanlike proceeding. On the other extreme, an undue anxiety to grasp the "horn of plenty" may result in the whole pack springing unscathed, when well within fair shot. Mallard and Wigeon (but never Teal) will sometimes give notice of their intention to spring by heading the wind and raising their heads. Of course, on such timely intimation the charge is despatched at once. But, in default of any such notice, it is no easy matter to correctly judge the distance while lying flat, and to seize the precise moment for placing the shot to best advantage. The middle course is, however, the safest, namely, to fire as soon as ever the forms of the fowl are distinctly distinguishable—say, between sixty and seventy yards, at which distance the pearly grey backs of the Mallards, and the white wing-patches of the old Wigeon drakes, as they sit up to flap their wings, can be clearly seen.

To return to the subject of the habits of the Ducks. In the absence of any such favourite diurnal resort as I have, perhaps somewhat lovingly, dilated upon, the bulk of both Mallard and Wigeon pass the day on the open sea, where the two species associate freely. Even at sea, Ducks usually have a distinct predilection in favour of some particular spot, to which they yearly resort, winter after winter. This is generally under the shelter of some point or headland, or, in the absence of these, of a reef of rock which affords some protection from the sea, where they prefer to sit close outside the line of the breakers. Hence, as the main bodies of Duck seldom enter harbour before dark, and leave it again before daybreak, and as on the sea they sit further in-shore than boats usually care to go, they are easily overlooked by those not acquainted with their habits. I remember an inland sportsman, inexperienced in the ways of Ducks, spending a whole week on the coast, literally almost within sight of thousands of them (Mallard and Wigeon mixed), and yet declare on his return that there "was not a Duck on the coast."

With regard to migration, both Mallard and Wigeon begin to arrive in this country in September. The Wigeon are the first to appear, their vanguard often reaching our coast during the first or second week of September. Towards the latter part of the month these are followed by small detachments of Mallard, and from that date onwards constant arrivals of both species keep occurring till the end of October, by which period the full complement of their winter numbers is made up.

In my opinion the whole, or, at any rate, the vast majority, of our sea-coast Mallards are foreigners. A few broods of native Mallards from the sand-links or immediate vicinity of the coast-line may perhaps join the foreign legions on their arrival here, but that is all. The inland, native-bred Mallards are extremely sedentary birds, and remain constantly, all the year round, in the neighbourhood of the moors and marshes where they were bred. These birds never, of their own choice, come down to the coast, or to the tidal estuaries.

WIGEON ON THE "SLAKE"—HARD FROST.

In very severe weather, when their regular haunts are all frozen or snowed up, they are obliged to have recourse to the open waters of the coast, but on the break-up of the frost at once return (within a few hours) to their inland home.

The return migration northwards takes place in March As early as the end of February, in mild seasons, we have evidence of the commencement of the great migratory movement, and its concluding stages are still perceptible in April, and even sometimes into May. But March is the month when the withdrawal of these ducks is in full operation, and, in average seasons, the great bulk of them leave our coasts during its concluding week.

Wigeon, on their first arrival, about mid-September, and

during the remainder of that month and a great part of October, are very apt to remain inside harbour throughout the day, instead of flying out to sea at dawn, as is their habit later on. During the period above mentioned, the Wigeon may be seen all day long floating lazily about the open water, or swimming round the edges of the mud-banks, plucking up the blades of the *Zostera marina.* Naturally, the punt-gunners take advantage of this habit, and during the first month of their sojourn on our coast a good many Wigeons fall victims to the big guns. Before the advent of November, however, their habits undergo an entire transformation. Whether they have learned wisdom in the bitter school of experience, or otherwise, it is rare to find these ducks about the oozes during the day in any considerable numbers after the 1st of November.

This phase in the character of Wigeon is rather remarkable, and appears at first sight to point to the conclusion that they are, by nature, diurnal in their habits, and that they are only driven to acquire night-feeding proclivities by the influence of man, and by considerations of safety. But, on further examination, this conclusion appears hardly to be borne out, though Wigeon are undoubtedly far more disposed to feed by day than are the Mallard. It must be remembered that, in their northern breeding grounds (whence they have newly returned), there is practically, during their sojourn there, no night at all. Even in Central Norway there is no darkness, and in their grand resorts in Lapland and corresponding latitudes, midnight is indistinguishable from noon. Consequently they then acquire promiscuous habits; and, like other Arctic voyagers, they eat when hungry and sleep when tired, without much regard to solar chronology. On first arrival here, the Wigeon, and especially the young birds, which now for the first time experience the regular alternations of light and darkness, continue the somewhat anomalous habits acquired in northern lands, where the summer sun never sets, or at least his light never dies out. In a few weeks, however, they adapt themselves to the altered conditions, and become absolutely nocturnal in their habits.

No doubt this change in their daily life is, to some extent, influenced by the greater amount of disturbance and persecution they undergo in our islands; but that this persecution is not the sole factor in producing the change is shown by the fact that their procedure is exactly the same in countries where they are but little or never disturbed at all. The writer has had opportunities of observing these ducks in various parts of the Spanish Peninsula, where their habits proved very similar to those noticed at home. Thus, on some of the large estuaries of North Portugal, where the natives had not then (at the time of my visit) thought of practising flight-shooting by night, the Wigeon for some weeks after their arrival spent the day on the grassy islands and sand-banks in the river. But by the middle of November they habitually went out to sea at the dawn. The only observable difference in their habits was the earliness and extreme regularity with which they flighted at dusk, owing to the accustomed absence of danger. Night after night they would appear within two minutes of the same hour—just before daylight had quite disappeared—and by my watch I could always time myself to leave the snipe-grounds and be in position for flight, with the certainty of having only a few minutes to wait for their appearance.

To return to our own coast, it is far from certain, even when one has the good luck, in winter, to fall in with a flight of Wigeon still lingering about the oozes during the day, that a shot will result. On the contrary, Wigeon are so much more suspicious of a gunning-punt than the Mallard, that, especially in mild, open weather, they will rarely permit of approach within any reasonable range. Wigeon, under such circumstances, begin to take notice of the presence of a punt as far away as 500 or 600 yards, and, as a precautionary measure, will paddle off into deep water; then, if pursued thither, will rise at perhaps 300 yards, and make straight for the open sea. In severe and frosty weather they are naturally less alert, and a shot may occasionally be obtained in daylight, though they are always a "kittlish" sort of fowl, and it is wise to fire as soon as ever a reasonable range has been obtained.

So far these notes have been confined to the diurnal habits of the birds under consideration. Now follows the night—by far their busiest and most animated period—and we will try to follow their fortunes under the moon. As the sun sinks below the land, and the gloom of the winter's night gathers around, there is commotion among those keen-eyed hosts which, since daybreak, have been rocking and tossing on the waves, or whiling away the hours on the sand-wastes. The sensation of hunger arouses them again to activity, and about an hour after dark—could one but see them!—they are rising in detachments, in little trips of two or three to a dozen or more, and speeding away separately through the darkness. Over the sea, and over the ranges of desolate sand-links, they hurry forward to the stretches of ooze and mud-flats within. All the day these wastes have been absolutely deserted, and, so far as ducks are concerned, quite lifeless. Now the dark skies resound with the sharp rustling of wings, and they circle lovingly over the broad expanse of succulent *Zostera*, gloating over the prospect of an aldermanic feast, and piping out their pretty resonant "whee-you." Then suddenly, from right beneath them, flashes a lurid gleam, and, as the report echoes across the waste, down falls poor "*Penelope*" flop on the mud. Away speed the survivors, but at point after point they meet with the same inhospitable reception. The "flighters" are out in force to-night, for the moon is well obscured by driving clouds, and the ducks are more easily discerned against their half-translucent masses. At last, in despair of finding a safe landing on the mud, down drop the Wigeon in the open water, and presently paddle cautiously inshore. But even then there is no absolute security. To the very outermost verge of the plains of rotten ooze, some hardy gunner, inspired to the tips of his toes with the predatory instinct, has managed to find a way by "plodging" down the course of some burn, whose shell-paved bed will just bear his weight. There he lies flat on his slimy couch; the armful of bent grass he has brought to rest on, already soaked by the rising tide, and the ooze and water slowly creeping into his sea-boots and all over him. Presently the little flotilla looms on the moonlit water in front of him. He will not move a muscle now, though the water rises inches around him, and, as the ducks draw inshore, he has the reward he sought. One by one his dog brings in the slain, and he departs homewards—satisfied. Verily it is hard and bitter work this flight-shooting on our British coasts in winter, and one can only admire the resolute pluck which alone can command success. But zeal is sometimes carried to excess, and, in my knowledge, the strongest constitutions have become mere wrecks from the long hours of wet and bitter cold spent on those wintry oozes.

The objects sought by the different ducks on their nocturnal excursions vary in almost each species. The Mallard, with its omnivorous tastes, is not confined to any single feeding ground, but speeds away on divers courses, some far inland to root about in potato fields, or search for stray grains in the stubble. Others make straight for some clover seeds or "hard corn" they wot of; while another, and far the largest contingent, remain on the tidal oozes to feed on the *Zostera*. These last show so strong a predilection for fresh water and its productions, that more of them will be shot by lying in wait about the places where small streams of fresh water from the land run down across the oozes than anywhere else.

Wigeon, on the other hand, feed almost exclusively on the green blades of the *Zostera marina* and other algæ and marine plants which grow in such profusion on the mud-flats, and seldom (so far as one can see in the dark) pass beyond the limits of the "full-sea mark." Neither fresh water, grain, nor potatoes have any attractions for them, the great oozy plains being their resort for the night. Here, as soon as the disturbances of the flight-shooters have ceased for the night, and each trip of fowl has at last managed to effect a secure landing on the salt-slakes, they get to work in earnest. An animated scene there must then be on the flats, under the rays of the moon, could one's eye but pierce her bright but deceptive light.

To come to terms with Wigeon, the best time is during the small hours of the morning, at a period when the tide happens to be from half to three-quarters flood. They have then fed, and will be found congregated about the edge of the rapidly disappearing mud-banks. The exact locality of the main bodies is not very difficult to make out by reason of their noisiness. The clear-toned "whee-you" of the drakes is audible to a considerable distance, and, on a nearer approach, the singular purring growl of the ducks is also distinctly heard. These single notes are incessant, but at intervals the whole pack burst out into a simultaneous chorus, which lasts perhaps half a minute, and then subsides. Wigeon, like some higher types of creation, are always noisy after a good dinner. In working up to Wigeon by night, it is absolutely essential to avoid going at all to windward, or they spring at once; they must also be kept full in the play of the moon on the water. Without this precaution they cannot be seen, even though within a dozen yards. With regard to seeing the fowl, to the writer it is only given to admire those who can detect Wigeon *on the mud* on a dark night, or even by such weak aid as starlight. This is, of course, a matter of eyesight almost as much as of practice; and personally I am always content if, under the most favourable conditions of smooth water and bright moonlight ahead, I can make them out as soon as within shot, or nearly so.

It occurs to me, on reading over the foregoing, that, between the flight-shooters at night and the punt-gunners in early morning, the ducks may appear to have but a poor chance. But this is not the case. Despite all the laborious efforts made to circumvent them, they are yet fully able to take care of themselves, and prove more than a match for the whole of us. On a great majority of nights they enjoy absolute peace, and at all times during low tide vast areas of the mud are inaccessible alike to man or punts. It must be remembered that not one night in a score is really favourable for night-punting. It is only during the period of moonlight each month that an attempt is possible, and, even then, adverse climatic conditions—such, for example, as a breeze of wind or heavy clouds obscuring the moon—may utterly preclude all chance of success. There are probably not a dozen nights throughout the whole winter, on an average, when all the conditions are even fairly suitable ; on some of these it may not be convenient to go out, and now the useless and mistaken legal restrictions have deprived us of the month of March—one of the best we had. The proportion of Wigeon killed annually is very small indeed, compared with the numbers of these birds. I have roughly estimated it, in different seasons, as varying from ten to fifteen per cent.

Before daybreak, or, at latest, on the first symptoms of dawn, both Mallard and Wigeon depart for the open sea to spend the day, and there, on our exposed coast, they are utterly inaccessible to man, whether by punt or sailing boat. I have seen (in pictures) a dinghey running into a page of sea literally strewn with dead and dying Mallards. This is not my experience in actual practice. The world is wide, and there may be spots where such feats are possible : I will believe when I have seen them performed. The game-ducks at sea I have always found fivefold more wary than the regular sea-ducks, and have never yet shot, or seen shot, a single Mallard or Wigeon from a sailing boat at sea. Often as I have run down on them, merely for the pleasure of seeing, say, a thousand ducks spring at once from the sea, I never knew them allow a boat to approach within shot, or, for that matter, within a quarter of a mile.

Next in importance to the two above-described ducks is the Teal ; but, in my experience, these game little fowi are but rarely met with on the north-east coast during the winter months. As early as August some appear on the salt water, and during September and early October they are plentiful enough, and right glad is the punter to welcome them. No sight is more gratifying than a flight of Teal, and no sound more pleasing to his ear than their low clucking note ; for, though usually unsuspicious of a punt, no fowl in existence is smarter or more game-like in springing, or requires more care and judgment to secure an effective shot. But in the month of October all these Teal appear to have passed on further south, or perhaps inland. Rarely are any seen on the salt water after that date, until their return northwards in March. Even in mild seasons it is quite exceptional to find Teal on this coast during the winter months. They are essentially lovers of fresh water. After punting for a week in January without seeing a single Teal, I have sprung half a dozen of them from a small fresh-water burn within a few hundred yards of the salt-slakes. Like all wildfowl that prefer fresh water and its productions, Teal are impatient of cold, and of the risks of having their feeding grounds closed by ice. Hence they move southward to avoid such dangers. Yet, once, during the intensely severe frost early in 1881, I fell in with six of these birds—all drakes—four of which I secured with a shot from a shoulder-gun.

The next four species of game-ducks I must dismiss in a very few words. The Pintail, Gadwall, Garganey, and Shoveller are never found to frequent the north-east coast in winter. Casual stragglers of these species may occur at irregular intervals, or on migration; but as habitual, or even fairly regular, visitants, the three first named are locally unknown. The Shovellers come every year to breed in certain localities; they arrive at the end of March, go straight to the pond where they intend to breed, nest in May, and as soon as their young can fly, at once depart for Southern Europe and Africa. Of the rest I have never myself met with a single example, and, with the exceptions above defined, they may be regarded (on the north-east coast) as simply non-existent. From what small personal acquaintance I have had with these ducks in other countries, they all appeared strongly addicted to fresh water, both by day and night. Our coast appears to lie north of their winter range.

The last of the surface-ducks regularly met with by the coast-gunner is the Sheld-Duck, a large and handsome species. They are resident with us, breeding in the sand-links, and bringing down their young into the water in July and August, and are more or less common all the year round. But in autumn their numbers are reinforced tenfold by arrivals from Scandinavia, and in winter I have seen as many as two hundred in a pack, though lesser numbers are more usual. Their favourite haunts are the "mussel-scaps," or stretches of mixed sand and mud, where various shell-fish abound, and a certain amount of sea-grass and green weed grows. Here they may often be seen by day moving actively about in search of crustacea, small shell-fish, and the like, keeping up the while a constant low sibilant note. The night is, however, their more regular feeding time, and by day they are commonly found passing away the hours at rest on the wide open sands already described as a favourite diurnal resort of the Mallards. However sleepy they may then appear, there is always a sentinel on duty, even if their company be only four or five strong, and it is not often possible to approach within shot, though I have occasionally done so, even in mild weather.

Like the rest of the surface ducks, the "Skells," when feeding afloat, frequently turn up vertically in order to reach the bottom.

SHELD-DUCKS.

MIDNIGHT ON THE OOZES.

A WINTER'S NIGHT IN A GUNNING-PUNT.

THE elements of hardship and uncertainty which are inherent to the pursuit of wildfowl have been alluded to; but it is *by night* that both are the most pronounced, and especially the former. Night-punting is certainly the most precarious of all British sports, and demands the greatest sacrifices of comfort. The boldest spirit may quail at the prospect of spending a winter's night aboard an open punt amidst the desolate creeks; and the keenest must stiffen his neck and harden his heart when it comes to turning out into a freezing atmosphere at midnight, just when the rest of the world are seeking their snug beds.

Moonlight is the primary essential of success—that is, the sport is limited to a possible eight or ten nights each month; but in practice there are seldom found more than a couple during each moon, when meteorological conditions are sufficiently favourable. On stormy nights one cannot go afloat, and even when the sea is calm the moon may be so overcast as to preclude any chance of discerning the fowl on the dark water. Then—cruellest disappointment of all—on those rare occasions when all the climatic conditions appear auspicious, and one sets out full of confidence, perhaps in the course of an hour or two the whole face of the heavens change, a breeze springs up, cloud-masses spin across the skies, obscuring the moon and "blackening" the waters—the game is up; the night's labour is lost, and nothing remains but to go home to bed—soaked, starved, and empty-handed.

Perhaps the readiest means available to draw a vivid picture of the vicissitudes of wildfowling by night is to narrate an ideal campaign as experienced during the month of February.

To begin with:—A continuous gale, blowing fresh from S. and W., for five whole days rendered all operations afloat impossible. Nothing could be done but "loaf," smoke, and watch a falling barometer. Weary days! The gale at length subsided, and the *sixth* evening offered fair promise of the patiently awaited opportunity: the moon, a few days past the full, shone brightly, and under her silvery rays the calm waters gleam clear and white. The tide would flow at 4 A.M., so an hour before midnight we launched the punt, and got under weigh with the first of the flood-tide. A couple of miles' paddling brought us to the outskirts of the ooze, and soon there was evidence of the presence of the *Anatidæ.* For miles along the dreary mud-flats rang out those inspiring notes, and this in a spot where *by day* not the ghost of a duck could be seen. Game-shooters—good sportsmen, too—who confine their wildfowling attempts to the hours of daylight, have before now returned from such a place in disgust, declaring "there was not a duck in the district." Nor is there, *by day*, but a change comes on the scene at night. Then, soon after dark, in their thousands the duck-tribe pour in from the sea, and by midnight the deserted oozes teem with wild profusion of bird-life.

At first we are only on the fringe of the feathered hosts—among the stragglers, single ducks, twos and threes. The tide being still low, these scattered birds were most difficult to discern, feeding among the loose stones and drift bunches of sea-weed which strewed the shore. They, in fact, usually detected us before we were aware of their proximity, which we only learned by the frantic quack, flutter, and splash as they sprang from the slob within a few yards. The main bodies are, during low tide, so straggled about as seldom to offer a tempting shot; over and over again we dimly discerned in the bright moonlight little bunches—fours, sixes, and eights—quietly swimming on the white water, or greedily dabbling on the ooze, and within half-gunshot. "Won't you take that lot, sir?" whispers my companion;

"there's nine together, all of a clump!" But I did not intend spending a winter's night at sea for a possible nine, and decided to hold on and await the more promising chance when the rising tide should have concentrated these scattered units into solid battalions. Moreover, except for a really heavy shot, I was reluctant to dispel the charm of the wild sounds and sights around us.

Seldom, indeed, is it possible to enjoy to such advantage the wilder scenes of nature, as on these occasions fall to the lot of the midnight fowler. All around him throng these creatures that, of all creation, most dread the human presence. Now the ooze and the moonlit water ahead are alive with sprightly, active forms, all feeding, playing and revelling in fancied security; overhead, the dark skies reverberate with the swish, swish of strong pinions as fresh "trips" pass and re-pass above. The variety of bird-notes and their musical intonations, to an appreciative ear, offer no small compensation for the hardships or discomforts of the situation, and also for a slight temporary restraint of the spirit of bloodthirstiness. From far and near along the flats resounds a continuous running refrain of blended individual voices. Then, at shortly-recurring intervals, the whole host join, for a few seconds, in one united chorus from thousands of throats; and this is followed by a few seconds of comparative silence.

Most kinds of wildfowl are distinguishable by their notes, and their concert serves to pass away those dark hours before the dawn. There is the low soliloquy of the Mallard drake, and the far noisier quacking of his wife; the strange half-"purr," half-growl of the Wigeon ducks, the long-drawn "whee-you" of the drakes. Even the gentle splashing of their bills as they dabble in the ooze is distinctly audible. All these and many others are well known—as familiar to the fowler as the bo'sun's pipe to a sailor. Then there is an almost infinite variety of notes—sharp, shrill whistles, low piping calls and undulated growls—which he knows proceed from the various wading birds; but to allocate each of these precisely requires more attention than the average fowler cares to bestow on these, to him, unimportant birds.

Presently there rings out through the darkness a loud, harsh note—a long-drawn, reverberating bark. That, too, he knows well; it is the call of the female Sheld-Duck—she, like the Mallard, being far more vociferous than her consort. Sometimes she winds up with half a dozen distinct *quacks*. The note of the drake is quite different—the peculiar, sibilant noise, half-squeak, half-whistle, before alluded to—usually quite low and gentle, but at times sharp and ringing. It is curious that the beak of the Sheld-Duck is tightly closed while the note is being uttered: the bird sometimes appears to be busy feeding at the very moment. Wigeon-drakes, on the other hand, open their beaks wide before commencing their pretty pipe, and close it during the note. Teal ducks quack not unlike Mallard, but lower, more hurriedly, and less defined, and these drakes also have a sibilant note. The young Teal when in packs in autumn keep up a constant low clucking chatter. Scaup appear silent—I never heard them speak—and Golden-eyes rarely, though they have a low, hoarse *quack*.

Noisiest of all his noisy race is the Curlew, the official sentinel of the wastes. His lung-power is simply terrific, and the vociferations of half a dozen, suddenly springing from a creek close by, fairly outrage the decencies of night, and spread an alarm for miles. I was amused to-night by overhearing my companion angrily muttering to himself that their conduct was "*parfectly scandalous!*" On a still, calm night such as this we could also distinctly hear the croaks and gabbling of the geese, sitting, full two miles away, on the open sea.

It was nearly 3 A.M. before the rising tide sufficiently covered the flats, and the chance we had awaited arrived. A mile or so beyond the spot where they had been feeding we came upon the now united assemblage of ducks, resting on the water of a sequestered little bay. By their notes we had little difficulty in making out their position, and presently drew up within sight of a fine flotilla under the rays of the moon. This was the critical moment. The slightest noise of man, boat, or gear—let an oar creak, or the setting-pole strike on a stone—and they are gone. No such ill-luck, however, befell us. When well within range the nearer stragglers began to "lift," and inside the same instant the roar of a thousand wings blended with the louder boom of the big gun, and a charge of No. 3 traversed their ranks. And now the silvery sea is strewn with dead, and shoving full speed ahead, the cripple-stopper is brought to bear on any that show signs of life. There are few cripples at night, and in less time than it takes to write this, all that we can see are secured, and a dozen or more of handsome Mallards and Wigeon will amply justify the prudence of the earlier hours of the night.

There still wanted over four hours to the dawn, when there was a possibility of again falling in with the ducks ere they take wing for the open sea. How to spend these hours is an ever-recurring problem in night-punting. To drop anchor and coil oneself up as snugly as is compatible with circumstances appears the easiest mode, but it is madness. So long as a man remains awake and in action, no cold will hurt; but to go to sleep in the night air is the height of folly, and sooner or later entails certain retribution. Suffice it to add, that a rather long shot, just before daybreak, increased the spoils of the night by another pair or two of Wigeon.

After a few hours' turn-in, and an unsuccessful campaign with the geese on the afternoon tide, we again went afloat at midnight. Again the ducks were there in hosts, but the conditions were changed. The sky was overcast, and the moon obscured by heavy drifting clouds, and, though several times close up to the coveted fowl, it was impossible in the darkness to make out their position, and we failed to obtain a shot. Once I was on the point of pulling trigger, but at the nick of time a glint of moonlight disclosed the fact that the dark objects ahead were *not* ducks, but some floes of drift ice, turning over, upwards and edgeways, in the tide-current, and whose moving outline closely resembled a nice "bunch" of fowl. Then, after eight of the coldest hours' patient effort, we returned to breakfast without a feather.

BRENT GEESE.

For coast-fowling *by night* (wrote Colonel Hawker), the Wigeon is like the fox for hunting—it shows the finest sport of anything in Great Britain. And the same remark applies to the Brent Goose for coast-fowling *by day*. The wildest and gamest, as well as the most numerous, of all our winter wildfowl, the Brent Geese, in hard weather and favourable seasons, afford right royal sport to the punt-gunner, and with this additional advantage, that the pursuit can be enjoyed exclusively during the hours of daylight. The following notes are the result of several years' tolerably close observation of the habits of these birds on the north-east coast, to which district the notes exclusively refer; for in other parts of the British coast their habits and migrations appear to differ to some extent from those herein described.

Though they are so abundant, and on many parts of our coasts the chief object of pursuit of the punter in mid-winter, yet on the north-east coast the Brents are quite the last to arrive, of all the host of migratory fowl which find a refuge on our shores from the rigours of the northern winter. The Wigeon appear in September, the Grey Geese and most of the Diving-ducks are all here before the close of October, but the Brents delay their arrival in force till the new year, or even later. The date of their arrival on this coast, as well as the numbers which visit us, are evidently regulated by the state of the weather at the different points of their range. Thus, while they have completed their domestic duties in Spitzbergen and Novaya Zemlya, and left those desolate regions by the end of August or early in September, yet they do not reach our coast *in force* for some four months afterwards. They appear to possess so strong a hyperborean

BRENT GEESE AMONG THE ICE

affection, that they will come no further south than is actually necessitated by their food requirements, being driven reluctantly southward, point by point, before the advancing line of the winter's ice. But in severe winters the congealing element carries stronghold after stronghold against them, and, as their last resources in the Baltic and in Denmark are closed up, they come here.

Though a small number of these Geese (fours and sixes) may generally be seen here about the end of October, and their numbers slightly increase as the winter advances, yet, as already stated, the great bulk do not arrive till after Christmas, and during January their numbers are being constantly reinforced by arrivals of "strangers," till their maximum numbers are reached, usually in February. In that month in severe seasons their numbers are often almost incredible. The "strangers" referred to are easily distinguishable on first arrival by their ignorance of local geography. It is often amusing to watch a big flight of them about daybreak, hungrily seeking the entrance to the harbour after their long journey. Each bird appears to have his own idea of the way in, to judge by the clamorous chorus they keep up; yet, after tacking off and on for half an hour, the whole pack will sometimes return to sea rather than trust themselves to fly over dry land, or into an ambush.

The following are the approximate dates of arrival of these Geese on the N.E. coast during ten years:—

1877. Middle of January.

1878. Early days of January; immense numbers in February.

1879. Second week in January.

1880. As early as December 7, 1879.

1881. January 15; very abundant in February.

1882, 1883, 1884. Mild seasons, with very few Geese; those which came, arrived in driblets in January.

1885. About a thousand arrived at Christmas, 1884; none came afterwards.

1886. Very few till the great snowstorm of March 1, when they arrived in unprecedented numbers, leaving again at end of month, as hereinafter described.

Brent Geese differ from the rest of the family in being exclusively marine in their haunts ; speaking generally, they spend the night at sea and the day on the tidal oozes, but never (like the Grey Geese) go inland to feed in the fields, or travel a single yard beyond high-water mark.

The habits of the Brents, if left unmolested to their natural bent, are as follows : After spending the night at sea, as the first streak of dawn appears in the east, they begin to think of breakfast. They rise from the sea, and, after ten minutes or so spent in preliminary evolutions, flying rapidly to and fro over the water, they head up, flight after flight, for their feeding grounds on the great zostera-covered mud-flats of the harbour or estuary they frequent. But the tide certainly affects their movements to some extent, and they prefer to come in on a flowing tide, especially at about one-quarter flood, if that period at all coincides with the daybreak. Of course this stage of the tide is only at intervals coincident with flight-time, and often very remote from it, and the Geese, like other creatures, have to submit to sublunary influences. Thus a few days before full moon, when the flood (on this coast) commences several hours before dawn, the Geese come in at that period—*i.e.*, in the dark. Such times therefore appear a good opportunity to bring the punt-gun to bear upon them, since they are somewhat dullish fowl in the dark—much more so, for example, than Duck or Wigeon—and we have them "inside" for an hour or so before daybreak. A good shot, however, is rarely obtained under these circumstances, for at such times the moon sets just at that particular hour. Hence the last hour before dawn is, perhaps, the darkest of the whole night, and the punter, therefore, is best in bed. Again, when the tide is falling at flight-time, the Geese come in on the ebb, though rather later (after daybreak). But since Brent Geese, though very active on their legs, have a strong objection to remain on the "dry" for more than an hour or two at a time, and are afraid to trust themselves in the harbour channels, a large number may be seen going out to sea about midday, returning on the flood in the afternoon. This reluctance to remain long out of the water is a factor in all their habits.

At frequent intervals they must go afloat to drink, splash about, and preen themselves. But I have perhaps written enough to show the sort of effect produced by the relative changes of time and tide, neap and spring, upon their movements. These, of necessity varying in each locality, must be ascertained for himself by every puntsman who wishes to be thoroughly conversant with their local habits and idiosyncrasies.

On alighting at their feeding grounds, the Geese at once commence greedily to pull up and devour the blades of the sea-grass, or *Zostera marina*; the whole black crowd advancing in the closest order over the green oozy mud, all heads down, except the sentries, of which an ample number are always discernible. With their small narrow bills, it takes a considerable time for a Brent to satisfy his appetite, and as the flood-tide covers the flats they still continue feeding as busily as ever, pulling up the sea-grass which grows within reach beneath them. Even at full tide, and in deep water, the Geese have no difficulty in obtaining abundance of food in the floating grass which is always carried off the flats by the tide. They never, however, dive for their food or otherwise, except when wounded, and, even then, they are very poor hands at diving.

In approaching a gaggle of Black Geese feeding afloat, during the exciting moments when the punt gradually draws near, a number of white spots are observed constantly to break their dark line, appearing for a moment, then vanishing. The effect is rather peculiar, and is caused by the Geese (as the water deepens) turning up vertically in the sea, paddling with their feet while their heads reach down to the waving grasses below. Their black fore-ends are thus submerged, and their white sterns protrude conspicuously.

After finishing their morning feed, about noon, the Geese are disposed to rest and spend the middle of the day floating about on the water, preening themselves; and, in mild weather, splashing about and chasing each other in sheer exuberance of spirit—very much analogous to what we would call a "flirtation." During this midday interlude they are very wide awake and absolutely unapproachable.

Probably no other creature in existence is wilder than the "gun-shy, punt-avoiding" Brent Goose. In open weather they are so utterly inaccessible that a whole season may go by without a single satisfactory shot being obtained, even where punters are numerous and the Geese thousands strong. Times without number one may "set" to them, but ere they are fairly in sight—ere one's eye can clearly distinguish their thin black line from the flat and featureless wastes—they are up! The distance at which these keen-eyed birds can discern so small and insignificant an object as a gunning-punt "end-on," is truly amazing.

Towards evening the Geese recommence feeding; and so intensely eager are they about sunset to utilize the few remaining minutes, that they then offer perhaps the most favourable chance to get within shot. The fortunes of many a long blank day have been completely altered, and patient hours of fruitless toil amply rewarded, by a splendid shot the last thing at dusk.

It is, however, in the hardest weather that the Brents afford right royal sport to those who then have the endurance to follow them. Then, when between tides the oozes and salt grasses are all congealed in the iron grip of the frost, the Geese are unable to get a bite during the ebb, and, as the tide flows over the mud, the quantities of drift ice which has been formed in shallow pools or in the stretches of "blown water" driving to and fro in the tide currents, the floes grinding and crashing against each other all over the flats, effectually interferes with their chance of a feed, and makes them less alive than usual to external dangers. Then (after a week or two of such weather) one begins to find the punt drawing nearer in upon them, and, at the short and deadly ranges which are then (and then only) attainable, one reaps an abundant reward for perhaps years of mild seasons, blank days, and numberless failures.

In approaching Geese (of any kind), as they do not rise vertically, spinning high off the sea at a single impulse, as Ducks do, but rise horizontally, going a yard or two before they are clear of the trajectory of a punt gun, one's fire should always be reserved till they fly. They are, however, much smarter than they appear in getting under way. Consequently the trigger must be struck at the same instant that the black pinions begin to spread. Not the eighth part of a second must elapse, or assuredly not a bird will be hit. The best shots at Geese are invariably made on the wing, by keeping a good elevation on the gun, and firing the instant they rise. The best time of all, when the heaviest shots are made, is on the break-up of long-protracted periods of frost and hard weather. At such times the Geese are often mixed with Mallard and Wigeon; in mild weather they keep separate.

Just at dark the whole host rise on wing together, and make for the open sea. In the morning they come in by companies and battalions, but at night they go out in one solid army; and a fine sight it is to witness their departure. The whole host, perhaps ten thousand strong, here massed in dense phalanxes, elsewhere in columns, tailing off into long skeins, V's, or rectilineal formations of every conceivable shape (but always with a certain formation)—out they go, some two hundred yards high, while their loud clanging "honk, honk!" and its running accompaniment of lower croaks and shrill bi-tones, resounds for miles around.

When much harassed and disturbed by punts during the day, Brents sometimes come into the harbour by night, especially during full moon; but they are fully aware of the danger of doing so, and of their own (comparatively) deficient power of vision during the dark. Hence they only trust themselves inside by night as a last resource, and when driven to it by hunger; they also, as a rule, take the precaution of returning to sea before the tide has flowed over the mud. They are seldom, therefore, obtained at night by punt-gunners, though a few fall victims to the flight-shooters.

Some of their habits, when wounded, are rather curious. After a shot on the mud, when the gunner goes ashore to retrieve his spoils, the winged Geese march away before him in a little herd. It is almost a ludicrous spectacle—the sportsman splashing and plodging half-way up his long sea-boots in the rotten, treacherous mud, with the little flock of

Geese waddling and croaking just in front of him, for all the world like an old henwife driving her brood to market. One by one, as they are overtaken, each bird lies flat down on the mud, stretching out his snake-like neck horizontally in a last hope of escaping detection. It is, moreover, astonishing how easy it is to overlook a Goose (which has fallen at a distance) when crouching thus. By taking advantage of some small hollow in the mud, they reduce themselves to much less bulk than one would imagine such large birds to be capable of; and so closely do they then resemble little bunches of drift weed, &c., that you may easily be searching all round, while your victim lies motionless, with his keen eye intent on you, within half a dozen yards.

The plumage of these birds is subject to some considerable variation. Generally speaking, it may be described thus: Head and neck black, sooty in tone, but yet glossy; back and wings dark slate-blue; the whole of the tail-coverts, above and below, pure white, causing the conspicuous "white sterns." It is the plumage of the under parts which varies most, this ranging from the palest grey (almost white in some birds) to quite dark colours. In some of the latter dark-complexioned individuals the breast is slate blue, almost as dark as the back; but in about one bird in fifty it is of a bright glossy-brown shade.

These dark-breasted birds are the exception, the vast majority being pale grey or dusky below, all more or less conspicuously barred, especially about the flanks. Then, on turning over a pile of Brents, one finds here and there a bird with pale edgings to the upper wing-coverts, forming a regular series of light bars across the wing. Such birds are understood to be the young of the year; if so, it is difficult to see why they should be, relatively, so scarce. The reverse is what one would expect, unless, indeed, it should prove that the young are less inclined to so far prolong their migrations, and that the bulk of our visitors are old birds. I should add that some of these bar-winged birds have the white neck-spot fully developed, more so than many of the plain-plumaged ones.

The migrations of these Geese do not appear to extend much beyond British waters. When shooting at Arcachon some years ago, I was told they were sometimes numerous there; but I saw none myself, nor did I meet with them during a winter's fowling on the north coasts of the Peninsula. Their departure northwards in spring depends to a great extent on the weather. In mild seasons it commences at the end of February, and continues during the whole of March and part of April. In 1883 (mild season) a part of the Geese were seen to leave as early as February 20; the rest in March. In the severe winter of 1879 my puntsman wrote me on March 25, "There is still a great many Geese left, but hardly so many as in February."

In 1881, after an unusually severe and prolific season, he wrote on March 17, "Part of the Geese left yesterday. They were seen going about east, mounting higher in the air as they went." (On the same date I read in the newspapers a notice of the partial re-opening of the Baltic navigation.) The rest of the Geese left that year on March 31. Some linger on our coast till April or even May; but, as above stated, the bulk are gone by the end of March. Yet, though they leave us in March, their breeding season in arctic latitudes does not commence till late in June. Their course, on leaving our shores, as observed by fishermen several miles out at sea, is invariably east, or a trifle to the south of east, pointing towards Denmark, which their instinct tells them is the most northerly point which the state of the ice at that period will permit. Thence they move on northwards by degrees. In 1886, on May 27 (just two months after they had left this coast), I happened to observe a migration of many thousands of Brent Geese when some fifteen miles off the coast of Norway. They were then bound due north, in all probability direct for Spitzbergen. Thus they appear in spring to follow northwards the retiring ice-edge as tenaciously as they retreat before it in autumn.

NOTE.—The Bernicle Geese (*Bernicla leucopsis*) I have never met with. Though numerous on the Solway and west coast, they are practically unknown on the east—whole decades elapsing between the chance visits of a few stragglers.

AMONG THE WILD GEESE.

A WINTER'S DAY IN A GUNNING-PUNT.

THE morning broke with one of those surprises to which in our "temperate" clime we are more or less accustomed. A sudden and heavy snowfall had occurred during the night. While men slept, all the familiar features of the landscape had disappeared, buried under the wintry mantle. Moreover, the feathery particles still continued to fall heavily, and with that steady persistency which bodes a "breeding-storm." How differently is such a phenomenon regarded! To the writer it is ever welcome, as presaging new campaigns among the wildfowl and fresh successes in the wild sports of the coast. The morning's post brought an invitation for a couple of days' covert-shooting to wind up the season (it was the middle of January), but this in the altered condition of things could not now be entertained for a moment. It is strange what an overpowering fascination the pursuit of wild-fowl has for its devotees. No other sport is so precarious, yet no one who has ever entered into its spirit, or been "bitten" by its enthusiasm, would dream of exchanging the chances of the gunning-punt, with all its risks, hardships, and uncertainties, for the most abundant game-shooting which the season will afford. That afternoon, therefore, I travelled down to the out-of-the-world corner of the county wherein are established my fowling quarters, more to "prospect" and arrange for further campaigns than in any great hopes of doing much so early in the season. On arrival the reports of fowl were satisfactory. Several hundreds of Geese had appeared within the last few days, and the evening was spent in discussions piscatorial and aucipial.

BRENT GEESE ON FEED.

To face page 200.

The object of the expedition being chiefly the Geese, which would then be snugly roosting on the rolling waves a mile or two outside the bar, nothing could be done that night or until the tide commenced to flow about six in the morning. At that hour the morning proved fine; the moon, only a few days past the full, shone brightly in the western heavens, and by her light we could dimly discern the desolate features of the broad estuary, extending far away inland, a dreary succession of dusky sand-banks and oozes, backed by the snowy outline of distant hills. The tide being now low, we had to

"THE LAST RESOURCE."—WINGED GOOSE TRYING TO HIDE ON THE BARE SAND.

launch the punt over some two hundred yards of sand and shingle—no easy matter with a craft some twenty-two feet long and so heavy as to require the full strength of my puntsman and myself to lift her on to the launching-carriage. Moreover, the sand was soft, and the wheels sank in places up to the axles, while ever and anon they ran against a half-hidden boulder. However, the morning was intensely cold—snow lying a foot deep down to high-water mark—so the hard work was not unwelcome, for it set the blood tingling through our veins. There is a certain strange weirdness about these dark hours just before dawn which is peculiarly impressive on the coast. The wind moans with melancholy cadence, there are dreary periods while clouds cross the moon, and the measured murmur of the dark wavelets on the shore has an "eerie" monotony. Of all Nature's creatures man, or rather that amphibious variety, or "sub-species," of our race which dwells on and lives by the sea, is perhaps the hardest working, and has the keenest struggle for existence. Already, at this early hour, the brown sails of the fishing fleet are disappearing in the gloom to seaward. They will be back with the produce of their "long lines" before noon, to get their fish to market that day, and the results will perhaps appear on the tables of the piscivorous, possibly hundreds of miles away, before night. That hoarse "clank, clank," resounding across the dark water, is also human; it proceeds from the small schooner which put in for shelter last night, and is now hauling taut her cable preparatory to getting away on her voyage by daylight. The only other sign of life is the weak little pipe of the Ring-Dotterels, running along the shore close to us in search of breakfast.

Our destination is the wide stretch of ooze where the *Zostera marina* and the samphire grow, and whither the Geese resort at daybreak, some three or four miles up the estuary. Our course at first lies across the harbour channel, where the tide-sheer knocks up a nasty sea, some icy cold sprays breaking on board of us. Just as daylight begins to break, my man descries some duck ahead, but not being myself endowed with crepuscular vision I fail to make them out. However, faith is still the essence of my creed, so we "flatten" on our chests, and after cautiously "setting" for some distance in the direction indicated they became visible—six Teal on the point of a sand spit. Unluckily we had forgotten to remove the handful of tow placed in the muzzle of the big gun to prevent her "drowning" as we crossed the rough water, and I didn't quite fancy the risk of firing thirty-two drachms of powder with so solid an obstruction in the barrel. However, Teal are the simplest of wildfowl, and as they sat well together, a shot from the large shoulder-gun stopped four out of the six—all drakes; lovely objects, with their exquisitely pencilled plumage and brightly contrasting colours. This was a good beginning, and a mile or so further up we observed a couple of Geese sitting on a dry sand-bank. They were evidently "pensioners," or pricked birds, so we decided to wait till the flowing tide should take us to them, when I killed the pair, right and left, with the cripple-stopper, as they rose off the sea within forty yards. This last acquisition, however, had cost us a considerable delay—over an hour—and during that time the main bodies of Geese had been passing in from the sea, filing off in long, black, gag-

SETTING TO GEESE.
"Right a bit . . . steady! That'll do . . . steady!"

gling skeins to the salt grasses ahead. And on our arrival on the edge of their feeding grounds, we truly appeared to have good reasons for abusing that unlucky pair of "pensioners," and our own folly in wasting a precious hour in securing them. For there, all congregated on the wide-stretching flat of slobby ooze, sat some thousand Geese, greedily guzzling on the succulent salt grass, while two creeks in the level mud, which appeared to converge on their position, were each occupied by a rival gunner.

How we anathematized our "ill-luck" (as bad judgment

or carelessness is usually called) needs not be told. Regrets and posthumous wisdom were alike of no avail, and nothing remained to us but to lie flat and watch the course of events. Gradually, foot by foot, as the flowing tide rose in the creeks, we watched our rivals pushing nearer and nearer to the black and clamorous phalanx before them. Presently they lay within a gunshot and a half, and their success appeared but a matter of moments. But a sudden change took place in the tide of fortune. All at once, and for no visible reason, the thousand pairs of dark pinions were spread, and with a sonorous roar the anserine host rose on wing. Directly towards us they shape their flight; close over the three prostrate punters passed their loudly gaggling columns, apparently quite unconscious of threatening danger, for in the open water just outside our unseen craft they splash down with wheeling flight and graceful evolution, describing in their descent a thousand eddying, opposing circles, concentric, eccentric, and elliptic. The position of the rival gunners was now reversed, for while the two "early birds" had to extricate themselves, stern first, from the creeks, we were in a position for immediate action. Luck had stepped in to help us where foresight failed! The Geese sat very scatteredly, so much so that while occupying acres of water they did not offer at any single point a dense mass on which to direct the stanchion-gun. In a few minutes we were close on their flank—already amongst the rearmost stragglers, and within range of the main line, when they again rose, suddenly and spontaneously as before. A shot as they rose would probably have secured four or five; but I prudently refrained, for they were only shifting their quarters, and almost immediately pitched again on the mud-edge, within a quarter of a mile. Once more we "set" in towards them: again we reach the fatal range, and ere they rise the big muzzle yawns within 100 yards of their dense ranks. Then the clamorous roar of their departure resounded: they had just risen clear of the mud, when the thunder of the stanchion-gun booms over the watery waste. Back rebounds the punt, and through the cloud of smoke we see the deadly result. Their line is broken, and the wide gap cut by ten oz. of B.B. is strewn with the spoils. Ten geese fall at once on the mud, another, hard hit, slants obliquely downwards, while from their retreating host a pair more of "droppers" turn over, and fall dead on the sea. Now follows a lively quarter of an hour with the cripples, and just as all are secured, down comes the snow again. Thick and fast it falls in blinding sheets, blotting out the sight of sea and sky, of geese and gunners. But the tide is now over; we have had enough; so with a fresh breeze, and the ebb in our favour, we set our sprit-sail and spin away homewards.

GREY GEESE.

There is no other division of our British wildfowl, the delineation of which I approach with so much incertitude, and conscious lack of precise knowledge, as the little genus defined by coast-gunners as "Grey" Geese. The genus consists of but four members, all closely resembling each other, hardly to be distinguished except when actually in one's hand, and all bearing a strong family likeness to their domesticated descendants of the farm-yard.

This uncertainty arises from no scarcity of the birds, or lack of opportunities for observing them. During six months out of the twelve the Grey Geese come almost daily under the observation of the punt-gunner on the coast; while, inland, they are the only geese met with. The Brents and Bernicles (which form the "Black Geese" division of fowlers), never quit the salt water: hence the long skeins of wild geese so often seen passing overland, all belong to the *Grey* division, but who can say to which species?

The difficulties which surround the problem of the specific identity of this group of birds arise neither from their scarcity, nor from any peculiar wariness on their part which is not common to all wild geese. It is rather the unpracticable, or inaccessible nature of their chosen haunts, and the resulting impossibility of obtaining a sufficient number of specimens at different periods, that leaves us so much in the dark as regards their specific distribution.

These remarks are undoubtedly at variance with the very confident and positive assertions of other writers on this subject; but they rest on experience, which always leads one to sift questions for oneself, and not to accept statements as facts merely because they are in print. In this study it is of the first importance to discriminate between the grain and the chaff of ornithological literature. Many writers have drawn conclusions from grounds which are far too slight, or insufficient, while others perpetuate error by simply transcribing the mistakes of their predecessors, or create a fresh set by substituting for missing facts a mere maze of guess-work. Such devices may serve the purpose of making their writers' books *appear* more "complete"; but it is infinitely preferable to be honest, to admit deficiencies in knowledge, and to indicate the points which remain in doubt. Those whose knowledge is the most complete will be the first to acknowledge the justice of these remarks.

Well, leaving for the present the doubtful ground of specific identity, the ordinary life-habits of the Grey Geese group are more easily diagnosed. They are among the earlier arrivals of our winter wildfowl. The middle of October is about the average date at which the Geese arrive in bulk; but the vanguard frequently appears in September, and exceptional occurrences even earlier. October is, however, the month when their V-shaped skeins are most often seen crossing the skies—each pack bound direct to some definite point, some resort they have perhaps frequented for centuries. We will accompany one of these skeins and follow their movements. First, as to the physical character and natural features of the locality which they seek for their winter abode. Grey Geese, unlike their somewhat distant relatives the Brents, claim to share the fruits of the earth with their arch-enemy, man. Grain is what they want, and, despite the most deeply-rooted fear and suspicion of our race, they will have it, and will frequent the arable lands so long as a stubble remains unploughed. After that, they will content themselves with the tender blades of clover or of meadow-grass; or perhaps will wing their way southwards to lands where a more bountiful nature, or a lazier race, dispense entirely with the plough. Feeding on the open corn-lands by day, their next desideratum is security by night. Inland they spend the hours of darkness in the centre of extensive pastures or an undisturbed lake or pool; but on the coast, if *Anser cinereus* could define his *beau ideal* of a "bedroom," it would be "ten square miles of dead-level sand, over which the highest spring-tides never flow."

Such are the two desiderata of these wary fowl, and to a locality affording both requirements, the Grey Geese come year after year, arriving in successive contingents till their full numbers are made up by about the end of October. Their normal habit, in common with the whole genus *Anser*, is to feed by day and to retire to sleep by night. But in October, when they first arrive, they find the fields full of workpeople, harvesters gathering in and leading the corn. Hence they are compelled temporarily to vary what is otherwise their normal disposition of time, in order to suit the exigencies of the moment. At that season, hundreds of Grey Geese may often be seen sitting huddled together during the day at their roosting places on the sand-flats. Now and then a detachment will rise, take a cruise inland, as though to reconnoitre the stubbles, and then return to their enforced meditations. But at dusk, as soon as the harvesters retire and the "coast is clear," away they speed in full force to gather in their share of the farmer's crop.

In several works on sport and natural history these birds have been described as *night-feeding* fowl—a mistake which has probably arisen from some such circumstances as those just described. The authors in question have arrived at a false conclusion, based on a half-truth. All Geese feed by day; and although at times compelled by extraneous circumstances to modify their normal life-habits, yet such variations are only temporary and exceptional. They are caused by the force of chance circumstances, and abandoned as soon as the causes cease to operate. In November, when the harvest is completed, and the fields comparatively deserted, the Geese no longer dream of nocturnal excursions. They then resume their temporarily disturbed habits, and as regularly as the sun rises, may be seen winging their way inland to the stubbles, and returning as regularly at dusk to spend the night on the sand-flats of the coast.

I should here mention that, never having had any shooting-ground to which these Geese resorted to feed, my experience

GREY GEESE ON THE SAND-BAR—"FULL SEA" (SPRING TIDES).

To face page 208.

of them is limited to the coast. Here, the only chance one can hope to obtain of securing them are on the sand-flats; and since the spots selected are seldom or never reached by the tide, this practically amounts to saying that they never offer a chance at all to the punt-gunner. I have never once seen the Grey Geese alight in the water or on tidal oozes where the flood-tide would set them afloat or enable a punt to approach within shot.

My brother Alfred and I have made several special efforts to obtain a few shots at these birds, in order to determine their species, but always in vain. On only two occasions have we been within measureable distance of success. On October 11th, A., who has paid special attention to the Grey Geese, and on whose observations many of these remarks are based, lay in his punt almost within shot of some 600; it was high tide, and the geese stood densely packed (no fowl sit closer) almost in the very wash of a heavy sea that was breaking along the shore. They were only separated from the stanchion-gun by a narrow sand-bar—180 yards across—and through which ran a deep, winding channel, intersecting the obstacle exactly at the point where the geese sat most thickly. This channel, with its shelving banks, appeared to afford an admirable means of access. But, alas! down that "gut" the tide rushed like a mill-race; once in its deep and surging torrent all control of the punt would be lost; boat and men would have been swept to certain destruction amidst the boiling breakers outside. A month later (November 11th) I observed eleven Grey Geese near the same spot. The tides being good, we awaited the flood, and set to them at "full sea"; but just as the advancing water, though a mere film, reached their toes, they took wing—in my limited experience these birds have always shown a decided antipathy to salt water. It being then within half an hour of dusk, we waited on, and soon heard the loud wild chorus of an approaching host. Presently they appeared—high in the clouds, and in two divisions. On nearing the sand-bar they lowered their flight close to the water, and amidst a perfect crash of clanging, stertorous voices, down they pitched, luckily on our side of the bar. At first we thought (being at some distance) that they had dropped in the water, and the tide being already on ebb, we went into action at once. Never had I felt such confidence in the prospect of at last getting a really good "rake" at these impracticable fowl. They sat massed more thickly than a battalion of Guards, and as the punt shot ahead the great birds towered tall as a herd of giraffes before us. I know my heart fairly bumped on the bottom-boards of the boat as momentarily we seemed to be getting on level terms with fowl that had so long defied us. Nearer still . . . . and then that horrid hissing sound told we were touching the sand. Only a yard or two nearer could she possibly float when, with another mighty outcry, up rose the geese and I fired. "How many down?" and I jumped to my knees, only to see with unspeakable vexation that the whole pack was scatheless. The great size and high carriage of these big Geese (they were sitting *dry*, not afloat as we had hoped) had completely deceived us, and instead of being well within 100 yards, the range, we found, had been nearly double. Then for six miles we "poled" homewards in silence, misery, and darkness.

To return to the habits of the Geese. With the first frosts of December, nearly all those which have arrived in October disappear from our coast. The departure of the "Harvest-Geese" on the approach of winter is one of the set phases in their life-histories. Then, in spring, they turn up again, and during March and April spend some six weeks or so here, on their way north. The local gunners hold that they leave us as soon as the stubbles are exhausted, and return in spring for the seed-corn; but it is probable that the weather, rather than food-supply, is the main cause of their departure. One other fact remains to be considered, as bearing on specific distribution:—namely, that during the hardest and most severe winters, in January and February, there is often to be seen an abundance of Grey Geese frequenting the same haunts, and living identically the same life as those already described, but which have departed. From an examination of the limited number of specimens I have had the opportunity of handling, all these *hard-weather*, or mid-winter geese appear to belong to the Bean and Pink-footed species (*Anser segetum* and *A. brachyrhynchus*), the latter species predominating. But to what species do the large spring-and-autumn birds belong? The Grey-lag is usually described as a scarce, and more or less casual visitant to our north-east coast, but perhaps the evidence rests on no very solid or sufficing basis. Mere market specimens I dismiss as utterly worthless, since, firstly, there is now-a-days no certainty as to whence they have come; and, secondly, as just described, a thousand geese may spend a month or two on our coast and depart without losing a single member of their mess, or leaving any "marketable" trace of their having been here at all.

Though we have failed to prove the case, I think that collateral evidence rather points to the probability of the double passage (spring and autumn) geese being all Grey-lags. The main breeding ground of that species is on the islands of the Norway coast, from Stavanger to the North Cape; they leave that country almost simultaneously about September 20th, but do not appear in their great winter resorts in Southern Europe till the middle of November. Where are they in the meantime? The interval coincides with the period at which we have the large geese above named on the north-east coast, which, moreover, lies directly on their line of route. Comparing the predilections of the respective species, the Bean and Pink-footed Geese are far more northerly breeders than the Grey-lag, neither nesting southward of the Arctic circle—I found the Pink-footed Goose breeding in Spitzbergen. Nor do either of these species pass nearly so far southward *in winter* as the Grey-lag, which nests in considerable numbers in Scotland and the Hebrides, and in winter is by far the most numerous of the geese that resort to the Spanish marismas. The Grey-lag, in short, is of far more temperate tastes than either of the other two species, which latter are the common winter wild geese of the British Islands. Their habits on this coast are identical with those of the autumnal geese already described. They pass the night roosting on the dry sand-flats, and by daylight pass inland to feed on grain, grass, and other vegetable substances. But they never, in my experience, pitch on water, mud, or ooze; or, in short, in any position in which a stanchion-gun can be brought to bear upon them.

The fourth species of the group is the White-fronted Goose (*Anser albifrons*); and it is a remarkable fact, and one strongly corroborative of the uncertainty which, as I hold, surrounds our knowledge of these birds, that of the merely trifling number of specimens which we have been able to secure, one has proved to belong to a *fifth* species, hitherto unknown in the British Isles. This is the Lesser White-fronted Goose (*Anser erythropus* of Linnæus), a bird which breeds on the Lapland fjelds, but appears to be of more easterly distribution in winter, frequenting the coasts of Greece and Egypt, especially the great lagoons of the Nile Delta, though I have recently heard of its occurrence in Spain at that season. This addition to the British avifauna was made by my brother Alfred, who shot a young male of this small Grey Goose on Fenham Flats, on the Northumbrian coast, September 16th, 1886, and the credit of the discovery is due to his correct identification. The specimen is now fully admitted to be an unquestionable example of *Anser erythropus,* the only one in existence killed in Great Britain.

The ordinary White-fronted Goose is more addicted to inland resorts than to the coast, and in the north-east is certainly not a common species, only occurring at intervals, and in small numbers. In the severe frost of January, 1881, a little string of eight passed close over the punt, so near that we could distinctly see their "barred waistcoats." I refrained from taking the chance with a shoulder-gun, judging, from their low flight, that they were about to pitch; but they passed right on, loudly cachinnating, never shifting brace, tack or sheet while in sight, and left us in the lurch—befooled once more!

I must now bring this chapter of doubts, surmises and uncertainties to an end. It is perhaps humiliating to admit, but the Grey Goose has proved *too many* for us. There are others among our winter wildfowl whose intense wariness all but sets at naught the most elaborate devices of the fowler; but with these there *will* occur an odd chance when, by some fortuitous combination of favouring circumstances, one may work a charge of B.B. among them. Not so with these geese; they come in hundreds, spend months at a time, are as regular in their habits as we are ourselves—but we *cannot get at them.* As already explained, they never alight where a punt can approach, or a punt-gun be brought to bear.

DIVING-DUCKS.

The Diving-ducks form a subdivision of what I have termed the Game-ducks, and are a well-defined and important little group, embracing several handsome species, and forming a regular (though minor) component in the spoils of the coast-gunner.

From the nature of their avocations, the Diving-ducks are almost entirely day-feeding fowl, as they require light for their subaqueous investigations. Those which prey on animal-food—living crustacea and other creatures which require catching—are exclusively diurnal in their habits; but one or two species, such as the Pochard, whose food consists of grass and vegetable substances, exhibit nocturnal proclivities. In the main, however, the Diving-ducks are of diurnal habits, and are met with during the day, inside the harbours or estuaries; in short, they occupy by day the situations then vacated by the nocturnal Game-ducks.

The presence of these fowl "inside" is, in winter, a distinctive feature in the sport of day-punting. The programme of the wildfowler at that season is practically limited as follows:—By night alone is it that he can hope to obtain a fair chance at the Mallard, Wigeon, &c. By day these are all at sea, and beyond his reach, and he has then only left to him the Geese, which, if the winter is mild and open, are pretty well inaccessible. Thus the chance of falling in with a company of Diving-ducks is a contingency that is ever welcome—on some days averting the calamity of an empty bag, and at more fortunate times adding a pleasing variety to the sport.

The most important species of this group of ducks, on the N.E. coast, are the Scaup and the Golden-eye. Both of them begin to arrive in this country during the month of October; but while the former is entirely restricted to the salt water and the immediate neighbourhood of the sea, the Golden-eyes distribute themselves throughout the country, being almost equally common on the inland lakes and rivers as on the tidal waters.

About mid-October, one may begin to look for the Golden-eyes, which arrive during the latter half of that month in small trips of from two or three to half a dozen birds. These, on first arrival, are quite tame and easily approached in a punt, before which they continue stupidly swimming away even when within fair shot. But a few weeks later, as soon as they have acquired experience of the dangers of the coast, Golden-eyes are among the wildest of all wildfowl; indeed, with the Mergansers they are perhaps the only birds which, on open water, it is wholly useless to try to approach in a gunning-punt. Golden-eyes, when on the coast, spend the night at sea, flying up in twos and threes into the estuaries at the dawn, and their haunts are the deep-water channels of the harbour, especially those with sandy or shingly bottoms, where they continue diving ceaselessly all day long. Their food consists of shrimps, small shell-fish and marine insects, besides, to a lesser degree, the sea-grass and other vegetable matter. This latter they often carry up from the bottom and eat at their leisure on the surface. I would not have thought them sufficiently agile to catch any of the true fishes, but one day last winter (Dec. 5th) while watching a Golden-eye busily diving among the ice on a small (inland) pool, I was surprised to see it capture several fish. Every third or fourth dive, it brought up a small silvery fish—sticklebacks probably—which it spent some time tugging at and chewing on the surface before finally swallowing. When feeding, the Golden-eyes are usually scattered about the "guts," and if for the sake of amusement, or in the absence of other game (for it is tolerably certain that no shot will be obtained), one tries to "set up" to a pair, their conduct is as follows:—They continue diving, first one then the other, often both under at once, and the punt draws nearer and nearer. There is no sign of alarm, and as our friends appear quite unconscious of our presence one begins to hope against hope. But it is all vanity. They are most deceptive birds, and at two gunshots' distance, without a sign of warning they are off—they seem to rise literally from mid-water, flying, as it were, from the very sea-bottom without tarrying a single instant on the surface. It is rare for many of these Ducks to be obtained by punters in winter (though in some seasons one sees them almost daily) except in *very* severe weather. Thus one January day, when the thermometer stood within a few degrees of zero, a bunch of about a dozen not only allowed my brother W. to approach within shot, but, the gun having missed fire, to replace the cap, when, being very near, the charge stopped nearly the whole lot, though several escaped by diving under the ice. These birds had probably been driven down to the coast by the severity of the frost from some moorland lough; in which situations the Golden-eyes remain comparatively tame and unsuspicious throughout the whole winter—the reverse of their behaviour on the coast.

The drakes of this species must take some years to acquire the handsome pied plumage of full maturity—perhaps three or four. One gets birds in all stages of the female or immature plumage, some with brown eyes, others hazel, and many light golden, and with the speculum, wing-coverts and neck-plumage in various degrees of development, but adult drakes are always extremely rare on the coast. We have only obtained one, and that so lately as Dec. 13th last—a lovely specimen in full mature plumage. It was a single bird, and so busy diving as to permit the punt to approach within shot of the small gun. Golden-eyes remain here till late in the spring. I have seen them in May, but that is not surprising, as in Norway they do not seek their breeding-spots, among the hill-lochs, till early in June.

The Scaup is another of the Diving-ducks which the punt-gunner is sure to meet with "inside" from time to time, though perhaps less often than the Golden-eyes. This, however, is not due to any relative scarcity of the Scaup, which some winters is quite as abundant, but is explained by some differences in their haunts and habits.

The favourite feeding-grounds of the Scaup is over rocks where sea-weed grows luxuriantly, and where they dive among the long, waving tangles in search of the various shell-fish and their spawn and the host of minute forms of marine life which abound in such places. Owing to this preference, their company is often confined all through the winter to certain localities—usually about the harbour entrance, or a rocky bay adjoining the open sea—hence they are less frequently met with than the Golden-eyes, which are scattered in odd pairs all over the sandy channels of the estuary. Moreover, such places as alluded to are not very

SCAUP DRAKE. (ADULT.)

accessible to punts; the water is too deep, and the long inward roll of the sea, even when smooth, is dangerous for these low-sided craft, to say nothing of the difficulty of using a big gun, when one moment half the fore-deck is buried in a great, oily, sloping swell, and the next the gun points heavenwards, far over the heads of the fowl. I have taken a punt, in broad daylight, within forty yards of nice packs of Scaup in such situations, but never could work a shot to advantage for the above reasons.

Besides the places where, as above indicated, the main bodies of the resident Scaup-ducks take up their winter quarters, one frequently meets with small bunches of half a dozen or so inside harbour, especially about the "scaps," or mussel-beds (whence probably their name), and even on the edge of the ooze, where they occasionally vary their shell-fish diet with a feed of sea-grass. They always, however, keep afloat, or nearly so; it is very seldom one sees a Scaup or Golden-eye go on to dry land, nor (on the coast) have I ever heard either species utter any note.

Scaup are the tamest of all the duck-tribe, and—exactly the reverse of the Golden-eye—they continue throughout the winter as tame and as easily approached as when they first arrive in October. On seeing a pack of them, one can shove the punt close in upon them, and then, if scattered, can wait securely till they arrange themselves nicely to receive the charge. Scaup are also among the toughest of birds and the most tenacious of life. At least half the cripples usually escape, and any that are captured alive it is all but impossible to kill. I have seen, when the bag was emptied on to the kitchen floor, a couple of Scaups, which had appeared as dead as door-nails, return to life and flutter vigorously round the room. Even when killed, however, they are of no value, being the strongest, nastiest, and most utterly uneatable ducks I ever tried.

The following extract from a note-book is illustrative of the two points just described—namely, the tameness and the toughness of these ducks. "January 5th: Early this morning came on four Scaup, feeding, half-afloat, in the dark; stopped three with a shot from 10-bore—only got two, a young drake and an old one. Later in day came on seven, very squandered. After getting within fifty yards, we lay by them for several minutes till they were all ranged in a line, when I fired and laid all seven on their backs—three apparently dead, four winged. Shoved in and commenced playing on latter with small gun, when one by one six of them disappeared, and though the sea was like glass, we saw not one of these again! We thus lost, at the moment, seven out of ten crippled Scaup, though three or four of these were picked up the same day, and nearly all within twenty-four hours."

Scaup also appear to be some years in attaining the full plumage of maturity. The young drakes, on first arrival in

October, show the full white front characteristic of the adult female. This they gradually lose during the autumn. In November, the whole head (including the white front) is spotted with black feathers, and by December the latter colour predominates. Young Scaup drakes shot early in January have a full black head, only the faintest traces of the white then remain, but the plumage is dullish black, only slightly glossed with green, and lacking the beautiful glossy reflections of greens and purples which distinguish the adult. The dark colour, moreover, only extends to the head and upper neck, the breast-plumage being still incom-

SCAUP—YOUNG DRAKE. (NOVEMBER.)

plete—merely mottled browns and greys, very different to the velvety purple-blacks and clean-cut waistcoat of maturity. The fine grey mantle is also but half developed, and in this stage, though the process of change appears irregular, and varies in different individuals, the bulk of the young Scaup drakes remain so long as we have opportunities of observing them on our coast.

The Pochard is now a very scarce bird on the N.E. coast. Thirty or forty years ago, according to the records of that period, and the recollection of old fowlers, it was an abundant species, and well known to both gunners and flight-shooters. There is no assignable reason for their withdrawal, but whatever the cause may be, the fact remains that at the present day the Pochard is now all but unknown. A chance straggler may now and then turn up in August, or while on migration, and a few years back I heard of two or three being obtained by flight-shooters in winter, but I have only once myself met with this duck on the N.E. coast. This was in January, during severe frost. It formed one of a little bunch of about a dozen ducks which were sitting on the point of a sandspit. We were in the act of " setting up " to them, when another gunner, concealed from our view by an intervening sand-bank, fired and killed six of them. Five were Scaups, and the sixth a Pochard in immature plumage.

MERGANSER DRAKE, showing form and carriage of Crest.

The Tufted Duck I have never met with on the north-east coast.

Another interesting and beautiful member of the duck genus which the wildfowler sees almost daily when afloat, is the Red-breasted Merganser. Exquisitely graceful in form and plumage, it is yet so wholly useless when killed, that no professional fowler would waste a charge of powder and shot over them. The Mergansers are, nevertheless, the most timid, wild, and utterly inaccessible of all the wild birds of the sea. So keen and alert is their vision, and so hateful the human race, that they will not, wittingly, allow the presence of a punt on the same square mile of sea as themselves; it is, in fact, often ludicrous to observe the immense distances at which their almost irrational timidity bids them decamp. Spending the night at sea, they enter the estuaries at dawn, and for the period of daylight succeed in setting at naught all the arts and stratagems of man—to them indeed, and to the Golden-eyes, belongs alone of all their watchful tribe the credit of outmanœuvring and nullifying the most elaborate devices of their arch-enemy. They systematically enter waters which are as free and open to punts as to themselves, remain there for their own purposes all day, and, evading every artifice to outwit them, leave again at night for the open sea, without losing the number of their mess. Of course, in punting year after year, a stray chance does turn up at intervals to work in a successful shot, but as a rule Mergansers and Golden-eyes are more than a match for the most skilful fowler that ever went afloat.

The only shots I have known at Mergansers from a punt have occurred either when they are caught sunning themselves round a bend in a curving sand-bank—this is a habit they often indulge in at mid-day, when a dozen may sometimes be seen basking together—or in a narrow "gut" where a punt can creep up unseen. They rarely, however, trust themselves in such dangerous spots, and if they should happen to find themselves hemmed in in a *cul de sac*, will attempt to dive back past the punt rather than fly over "dry" land (or what Mergansers may regard as such). They feed entirely on shrimps and small fish, and are quite uneatable. There are, however, few more beautiful objects than a newly-killed Merganser drake. As he lies on the fore-deck—the weird, half-uncanny expression in his blood-red eye still undimmed; the slim, snake-like neck and glossy head, adorned with its long double crest—one-half standing straight out backwards, like the "toppin" of a Peewit, the other pointing downwards towards the back (*not pendent*, as invariably represented in books); then the lovely but evanescent salmon hues which tinge his breast—all these points, together with the bold and brightly contrasted plumage, combine to form as beautiful an object as any that Nature has produced.

Towards the end of February, Mergansers, and other ducks, begin to move northwards, and at that period we often observe, on the N.E. coast, small parties of this and other species putting into our harbours for rest and food, preparatory to continuing their journey by stages. On March 1st, 1881, I was cautiously following six Mergansers at low tide in the punt, when, on rounding a turn in the sandbank, they all landed. It was a most interesting sight to watch their sprightly graceful carriage as, half upright (*i.e.* at an angle of about 45°), they ran up the sloping sand in most active style—very different from the waddling gait of most of the Diving-ducks, some of which appear almost unable to stand at all. The Scaup and Scoters are seldom seen ashore, but when driven to it, sit awkwardly with their great splay feet turned inwards in most ungainly style. The further back a bird's feet are placed, the more upright it necessarily stands. Thus the Cormorant and Merganser sit as described—about half upright (45°) : the Sea-ducks rather more horizontally, and the Wigeon is actually horizontal. Guillemots and Grebes, whose legs are practically terminal members, sit bolt upright, while, so far as I have been able to see, the Colymbi are unable to stand at all.

To return to my six Mergansers : they were evidently paired, for after landing they separated into three pairs of *fiancés*, one of which I shot, and thereby, perhaps, saved them from future remorse and recrimination ! These birds roost on the sea, and are exclusively marine in their haunts ; I have never, in winter, seen them away from the salt water, whereas their congener, the Goosander, though not uncommon inland, rarely visits the tidal waters. Their haunts are the freshwater streams and large rivers, where they feed on trout. I only once remember seeing what I took to be three Goosanders on the coast, but Mr. Crawhall has shot one coming in from sea at the morning flight.

There remain two other members of the Mergus tribe, which are invariably mentioned by writers on wildfowl ; but neither of which I have ever seen alive—namely, the Smew and the Hooded Merganser. The former can only be regarded as an extremely rare winter visitant, which has, perhaps, never occurred *on the coast*, and as for the Hooded Merganser, I may state my private opinion (despite the very circumstantial descriptions of it given by certain writers, which I regard as mere guess-work), that it has never occurred in Britain at all. None of the evidence of its alleged occurrences is direct, or sufficiently satisfactory to establish a scientific fact. That the Hooded Merganser does visit Ireland has recently been demonstrated by the tangible proof of three specimens shot in that country by Sir R. P. Gallwey. Being a North American species, Ireland is where one would most naturally look for its cis-Atlantic appearance; but, so far as England or Scotland is concerned, the Hooded Merganser might well be expunged from the list of feathered inhabitants.*

* The above remarks may appear too sweeping; but I prefer to let them stand, and for a statement of the facts respecting the occurrences of this and other reputed British birds, would refer my readers to "Yarrell" (4th ed.), where all the evidence is set out with judicial impartiality—as regards the Hooded Merganser, at vol. iv., p. 510.

SEA-DUCKS.

In describing the habitats and the natural economy of the various members of the duck-tribe, and of other birds whose haunts are remote from the arm-chair of the Natural-historian, and which cannot be observed during his after-breakfast ramble, it has been customary to write somewhat vaguely. Thus, on looking over some books on Natural History, the haunts of any given species of duck will be stated to be "arms of the sea, rivers, lakes, and marshes." The next species will be described as frequenting "marshes, lakes, rivers, and arms of the sea," or, perhaps, such expressions as "creeks, pools, and moist situations" may be substituted. There is obviously an abundant scope in such descriptions for synonymic verbosity and neatly-turned paraphrase; but when all that is written is "boiled down," it amounts to little more than platitude. Every schoolboy knows that ducks swim, and require water to swim in; and this, despite all the redundant verbiage employed, is about the sum and substance of the information that can be extracted from three-fourths of the *popular* works on this subject. Endeavour to ascertain from them any special feature—consult them with a view to confirming personal observations or ideas—they are silent.

Yet the duck-tribe vary in their haunts and habits, as between one species and another, quite as much as any other family of the feathered race. The poverty of description simply arises from their being less understood. These variations can only be accurately observed, or, at any rate, are observed to the best advantage, by those who are sufficiently enthusiastic to follow the regular sport of wildfowling afloat, and who alone enjoy the opportunity of becoming acquainted with these wild creatures in their bleak and desolate haunts. Hitherto, unfortunately, but few of those who have become enamoured of this sport have paid much attention to, or, at any rate, displayed much knowledge of natural history.

In the foregoing articles I have described, as far as my opportunities of observation permit, the habits of the different groups and species of ducks and geese which are comprised under the term of wildfowl, as they come under the notice of the coast-gunner. But the British ducks are a numerous family, and there remains a section which the punt-gunner never meets with, and of whose existence he might remain wholly unaware, so long as he confined his operations to the punt and the sheltered waters which alone are navigable by these craft. The group of ducks to which I refer do *not* frequent "rivers, lakes, or arms of the sea"; they do not enter harbours or creeks, but their haunts are exclusively on the open sea itself. The sea-ducks comprise the Scoter and the Velvet Scoter, the Eider, the Long-tailed Duck, and, to a less extent, the Scaup.

The last-named, as already described, is not infrequently met with inside harbours, where they go to feed on mussels and such-like shell-fish. Still they are mainly sea-ducks, and a favourite resort is a rock-bound, weed-covered bay on the open coast. Under the shelter of a long black reef or scar, or within a narrow bay on a rocky coast, a company of Scaup will take up their quarters for the whole winter, and seldom leave the spot, unless disturbed by man, or driven out by a heavy sea. From their unsuspicious nature, it is not difficult to approach them, and a pretty sight it is to watch a company of them in such a place all busily engaged on their every-day employment. The nearly white back of an old drake contrasts prettily with the dark weed-covered reef along which he cruises, ever and anon diving close under the rocks to study conchology among the waving fronds and sea-tangles which grow beneath him.

The Eider resembles the Scaup in many of its habits, and both ducks are intimately acquainted with the local geography of the sea-bottom: all its depths for miles, and the position of every submerged reef and shallow are well known to them. But while the Scaup contents himself with the smaller shell-fish and crustacea, the Eider, with his strong, hooked beak, can crush and devour dog-crabs nearly as broad across as one's fist: from the gullet of an Eider drake I have shaken out three or four big crabs, on holding him up by the legs.

Eiders are specially fond of going ashore to sun themselves on the edge of a reef or rocky island. In such positions, among the black rocks, one would imagine an old Eider drake would be a very conspicuous object; but it is not so. It is surprisingly easy to sail past a dozen of them unperceived, so precisely does their bold black and white plumage harmonize with the broken water, and with the great balls of foam which are driven up on to the rocks by the wind and sea. Eiders, or, as they are locally called, "Culvers," are quite common on parts of the N.E. coast; but Northumberland has the honour of being the only English county where they remain to breed. Their nests are placed both on the rocks and among the bent grass along the sand-links, and contain five green eggs. During winter, the sexes are often found congregated separately, but in that case there are usually to be seen a few precocious females among the packs of drakes. In the month of March, as the nesting season approaches, Eiders are apt to draw into the harbours and sand-flats—places they never frequent in winter.

Another handsome bird which spends the winter with us is the Long-tailed Duck. These also get their living by diving, but in a very different manner to the two species just described, whose food, as stated, consists of shell-fish, thus restricting them to places where the bottom is rocky and of no great depth. Hence the Eiders and Scaups are usually met with close inshore, or, if found diving at a distance from land, the fact will be explained by the existence of a submarine reef. In no case do they dive where the depth exceeds, perhaps, two or three fathoms. Their food is exclusively *on the sea-bottom*; but that of the Long-tail is in mid-water; that is to say, the latter bird does not require to reach the bottom at all, its food consisting of animalculæ and other minute creatures which swim at all depths. Hence the Long-tails, and, in a less degree, the Scoters, are able, like the Guillemots and Razor-bills, to live in sea of any depth, and can often be seen busily diving several miles out from land. On examining one of these ducks when newly killed, it is impossible not to be struck with the difference in the form of their small bills when compared with those of the rest of the sea-diving ducks. The latter are heavy and swollen, broadening out towards the tip—regular shovels in fact. The bill of the Long-tail, on the contrary, is small, narrow and delicate, narrowing to the tip, but strongly pectinated, or furnished with a comb-like process admirably adapted for sifting animalculæ, &c., from the sea-water, but not for catching crabs, &c., as the rest of the sea-ducks do. At the same time, these ducks are quite capable of subsisting by bottom-feeding, and are often to be seen diving in quite shallow water near the shore, where they feed on small shell-fish. At one point on this coast, where the depth rarely exceeds a fathom or two, over a shingly bottom, a company of Long-tails are nearly always to be found, as well as a few of the somewhat similarly formed Golden-eye.

The Long-tailed Ducks are rather late in arriving—often not till November, and disappear during the early days of April—sometimes simultaneously with the advent of the Terns, about April 8th. In their build these ducks are heavy and thick-set, like the rest of the diving-ducks; not long and slim, as most illustrations of them appear to convey. Though the females are always plain and sombrely clad, an old drake, when newly killed, with his chaste and harmonious plumage, is a strikingly handsome object.

From the nature of their haunts it is impossible to get at any of these sea-ducks in a gunning-punt—these craft being only available in smooth or land-locked waters. Outside harbour, however, some most enjoyable days may be spent, in a well-frequented locality, cruising about among the fowl. If the day is fine, with a good, steady land-breeze off-shore, a few fair shots may now and then be got at sea-

Q 2

ducks with a shoulder-gun; but any one who has tried it knows how rarely ducks of any kind will allow approach in a sailing boat at sea. Still there is, quite independently of killing or sport, a very great charm in shooting under sail. Not only is there an opportunity of observing many wild and interesting fowl, but the sensation of spinning along in one of the fast-sailing cobles of the N.E. coast, as she walks through the seas with a huge wale of hissing angry waters rising in a menacing slope high above her lee gunwale, is in itself most exhilarating. Presently the look-out descries fowl. "Luff!" he whispers, "Covies (*i.e.* Scaup) bearing the South Beacon!" and in a moment the boat is beating up to windward. "Keep your luff!"—why, we can hardly keep our seats as the stout coble thrashes through the seas, close hauled to the windward and the flying scud, to say nothing if an occasional bucketful of *green water* drives athwart her. Nothing short of oilskins will avail to keep one dry as she labours ahead, full and bye, and with the leach of her big brown sail temporarily stiffened with the boathook. At last, when the weather-gauge has been attained, up goes the helm, and, with a flowing sheet, we run in on the pack of Scaup gently rising and falling on the swell. But even Scaup, tame as they are, won't allow a big coble to run right over them at sea, and long before we are in shot one sees the little white jets of spray flying up here and there among the ducks as, one by one, they rise heavily and get under way. Poor birds! They have done their best to secure safety; but instinct, or, at any rate, reasoning power, lacks a little at this point. As they steam away full speed in a straggling line to windward, they sometimes fail to observe that the coble's course is once more altered. Under a lee helm she flies up again into the wind, and, with her gathered way, is scudding right into their "line of communications." Moreover, if she has been well handled, she has, under favourable circumstances, perhaps a less distance to traverse than the birds, and this (unless, as of course often happens, they change their course) will bring the fowl right across her bows—indeed, sometimes right over them. It is, of course, a very old manœuvre—cutting out ducks by a "luff," as they fly to windward—but still it is a very pretty one, and affords some exciting moments.

The behaviour of all the sea-ducks when approached by a boat is usually the same, *i.e.*, they rise wide, but may now and then, under favourable conditions and in a steady breeze, be cut out as described. No duck of any sort whatever would ever dream of attempting to escape by *diving*. I have been told that the Long-tails occasionally do so; but I never saw any approach to it myself, and fancy that in such cases either the birds had not realized the presence of danger and were simply diving for food, or that Grebes, Tysties, or other birds had been mistaken for ducks. It *may* be that the Long-tails will, under certain circumstances, attempt that means of evading pursuit, but such conduct certainly appears to me very improbable and entirely foreign to what I have observed of their natural disposition. What Mr. Folkard is referring to (and 'Wild-fowler' copies his mistakes) when he describes the duck-tribe (and especially such species as Shoveller and Sheld-Duck, which are strictly surface-ducks) as habitually resorting to diving as a means of escape from danger, is quite incomprehensible—but so, I must add, are many statements in both those books. Mr. Folkard speaks of chasing Shovellers about with a rowing boat (p. 260): as well, in truth, might their pursuit be essayed with a four-in-hand!

Although, however, the sea-ducks invariably use their wings in preference to their legs in order to keep clear of danger, yet, *when winged*, so proficient are they at diving, that it is all but hopeless to attempt to capture them. Half a dozen Scaup, Scoter, or Long-tails may fall to a shot, but, except the dead, not one will ever be seen again save by a mere chance. They appear to dive straight as they fall, and nothing more than the point of a bill will again appear above water till danger is past. Winged Eiders, as a rule, can be followed up and occasionally secured if the sea is quite smooth. They have hardly the same power of holding themselves just under the surface, and, being so large, are more easily seen when they reappear. They rely on the immense distances they can traverse under water, and generally with good reason. But as for the others, give it up, gather the dead and go on to try for fresh ones, for the winged you will never get.

More numerous than all the above-mentioned species together is the Common Scoter or Black Duck, which comes to the coast in swarms. The open sea is their home; they may be met with diving in twenty fathoms several miles from land, or at other times close along the shore, feeding off the reefs which fringe the coast line. Inside harbour they never go—I only once remember seeing one there—a single bird shot by my brother, but it had probably been "pricked."

Scoters are resident here all the year round. Throughout the whole summer flocks of these ducks still frequent their winter haunts off the coast, though they are a northern breeding species, none ever nesting in England. These summer birds are all immature, from which it appears clear both that this species requires, at least, two years to attain full maturity, and also that they do not breed till that stage is attained. Many young drakes (of the second year) shot in February are half changed to full black plumage.

Abundant as these ducks are they afford little or no sport; being equally distasteful both to eye and palate, they offer no reason or excuse for pursuit after a few have been obtained. In the last-named particular they are, it is true, no worse than the rest of the sea-ducks; but the others have at least the charm of beautiful plumage, which is wanting in these "ugly ducklings." The Velvet Scoter is a larger and handsome species, the jet-black plumage of the old drakes being peculiarly rich and glossy, and is easily distinguishable at any distance by the broad white speculum on the wings, closely resembling an old Blackcock, if one could imagine such a bird far out at sea. They are far less numerous than the Black Scoters, though a small company or two of half a dozen birds each may generally be met with in the same localities as these ducks. The Velvet Scoters, and the winter contingent of the smaller species, both arrive here in October and withdraw at the end of March. Beyond seeing them at sea, I have not had much opportunity of observing the habits of the Velvet Duck.

Such are the regular "Sea-ducks"; but in the course of a day's cruise, one often falls in with other kinds. Thus one often sees the Sheld-Ducks spending the day at rest on the waves—sometimes seventy or eighty strong—and is almost sure to come across the local stock of Mallard and Wigeon sitting along shore and close in to the line of breakers.

The pleasure of shooting under canvas is further enhanced by the constant opportunities it affords of observing various wild creatures other than the *Anatidæ*. During mid-winter we have the Little Auks from Spitzbergen, and the pretty coral-footed Sea-pigeon, or Tystie from Shetland. These are replaced as spring approaches by the arrival (in March) of the Puffins. The two first named are more or less oceanic in their resorts, but the Common Guillemot and Razor-bill are ubiquitous and hardly take the trouble to get out of the coble's way. Then there are the large Divers (Colymbi)—I have shot all the three species—and four kinds of Grebe may be met with: one of these, however (the Eared Grebe) is decidedly rare. Lastly, there are the seals—weird, uncouth amphibians. As they silently gaze on one from the sea with their great mild eyes, they verily impress one rather as the ghostly relics of some long-past Arctic epoch than as contemporary denizens of British seas! Seals still breed at a few spots on the north-east coast, bringing out their young in November on some remote little islet or "skeir" of rock, just awash at full tide. On these islets half a dozen or more of their ungainly forms may sometimes be seen basking in the wintry sun, while, hard-by, stand the gaunt upright figures of the Cormorants.

The serious drawback to the pursuit of wildfowl at sea, is the constant risk of being caught in a sudden gale, perhaps when several miles from shelter. This contingency, which is ever impending, not unfrequently bursts upon one without notice, and a most unpleasant experience it is to undergo. Seaworthy as the northern cobles are, they, and in fact all open boats, are unmercifully wet in a gale at sea. Many a day which opens auspiciously with a bright morning, a calm sea, and a fine land-breeze, ends in a wet and miserable scuttle back to harbour in the teeth of a nor'-easter, with three reefs in, and an angry sea playing upon us in pelting cascades like the jets of a fire brigade.

WADERS, DIVERS, AND OTHER FOWL.

ONE of the charms of coast-shooting is the variety of the game. The wildfowler at any rate enjoys seeing (though he may not obtain) a far greater variety of game than is the case in any other branch of sport—incomparably greater than on the fells, fields, or woodlands. In punt-gunning, unfortunately, it is simply impossible, at once, fully to combine the characters of both sportsman and naturalist—at least, as regards the securing of specimens. One or the other character must be predominant, or failure and disappointment are tolerably sure to be the result of an attempt to sit on two stools. One of the first essentials of successful fowling is *quiet*. Cruise about unheard and unseen as far as possible; disturb nothing, not even a gull; and never fire a shot till at length you have the grand opportunity you have for hours been manœuvring to secure well under your gaping muzzle. The man who is perpetually pop-popping at Plovers or single birds in hopes of some day securing a "rare specimen," will not only utterly ruin his own chance of getting a good shot, but that of every one else in his neighbourhood.

From a purely ornithological point of view, this is a matter of regret, especially when one remembers the malignant persistency with which these stray chances at out-of-the-way fowl invariably turn up at the wrong moment. One may spend days, weeks, without seeing a creature beyond those telescopic-eyed geese and impracticable Wigeon. Then, just as one is "flattened" to what looks like being a favourable chance of a big shot, there floats past within half-gunshot a single Grebe, or a pair of ducks one does not recognize, either of which *might* prove of exceptional interest. But to fire at them would be madness; your puntsman would most likely resign office at once in disgust, and a possible prize drifts out of sight, never to appear again.

With the writer the point has never been in doubt, and however much the necessity may be regretted, the pursuit of sport must be paramount. Otherwise, no one would be found willing to undergo the hard labour and the long cold hours merely on the chance of getting, once in a lifetime, a really "rare bird." This, too, is on the hypothesis that such a creature exists, a proposition which (except relatively) is generally untenable.

The chief and most interesting fowl met with, beyond the ducks and geese described in the foregoing chapters, are the

GROUP OF SMALL WADERS.

three classes known as Divers, Grebes, and Waders. Of these, the latter are by far the most numerous, and are, moreover, creatures of such exquisite grace of form, of plumage, and of motion, as infallibly to engage one's interest and attention. The waders are the earliest migrants to reach our coasts in summer—the vanguard of the feathered hosts from northern lands. During the spring and summer months the stretches of tidal ooze and sand lie dreary and almost lifeless. Visit a great estuary in June or July—you may ramble for miles around its shores, the scene of the winter's exploits (and failures), and call to mind the noisy flights of wildfowl seen, and the exciting moments enjoyed, on these very spots in January and February. Now, there is hardly a living creature to enliven the dreary monotony of the wastes. Now and again the glint of a sea-gull's wing, or perhaps a brood of young Sheld-Ducks—that is all one sees in several hours' ramble. But in August a change occurs. In a few days the shores are once more enlivened by the cheery sights and cries of a profuse bird-life.

As early as the end of July the Whimbrels and the Arctic Skuas appear. Both these birds breed in Shetland, and have not far to come. Distance, however, is a mere nothing to these cosmopolitan wanderers. It is an element which is practically eliminated from their reckonings by the trim build and wide pinions of even the tiniest waders. Thus the Shetlanders are hardly here ere there pours in, close behind them, a perfect flood of travellers from the highest latitudes and most remote spots in the known world—aye! and beyond it too. Purple Sandpipers from Spitzbergen and Turnstones from Nordland throng the rocks; Godwits, Knots, Grey Plovers, and Sanderlings from Asiatic *tundras*; Greenshanks, Ruffs, and Whimbrels from various points between Sutherland and Siberia, and a host of cognate birds from the morasses of Lapland and the Norwegian fjeld suddenly populate our shores. In September the Curlew-Sandpiper arrives—who can say from where?

Many of these birds have come to spend the whole winter on our coasts; but a large section only appear here in transit, passing on southwards at once, not to reappear till their return journey northward in the following spring. This latter group comprises those species which, seeking their food largely among fresh water and its productions, are dependent on mild, warm weather. They are impatient of cold, and must always keep well to the southward of the risk of frost—which to them implies starvation.

Within this category fall the Whimbrels, Greenshanks, Common and Curlew-Sandpipers, the Stints, and the Ruff. The through-transit of these birds continues during the months of August and September; but it is probable that no individual bird spends more than a few days on our coast, the period being occupied by the continuous succession of fresh arrivals and departures, lasting till the whole bird-population of these species has completed its passage. The Whimbrels, while here, frequent both the mud-flats and sandy shores, and are also much addicted to the rock-pools and weed-covered rocks, where they feed on small dog-crabs and other shell-fish. About the 25th of August, the Greenshanks appear, always a scarce bird, though one or two seem every year, at the same date, to return with remarkable persistency to the same pool or stream, and this, too, although the particular individuals have been shot each year on arrival. The young of the Common Sandpiper comes down from the moors at the end of July, but only passes a few days on the rocky shores before proceeding onwards for the south. The larger Curlew-Sandpiper is later in arriving, not being due till September, and moving on within a few days. By September 20th they, and the Greenshanks, have passed right on and are gone. The Ruff is also a scarce bird. On August 23rd (my brother writes), we were rowing up a bight in the slakes, when I chanced to see a bird stretch its wing and gently close it again. It was sitting among grey whelk-covered shingle, and, though close at hand, I could not make it out, so fired at the place. Nothing flew away. On coming up, two young Knots and a Reeve lay dead.

Such, roughly described, are the birds which every summer pass southwards along our shores to the aggregate of many millions. Some of each and all the species named, may be secured any August by those who know where to look for them, for almost every kind has some special resort to which it is more or less confined. But, in order to complete the list of this group, I must mention also the following allied birds, which also pass south at about the same period, but so scarcely or irregularly that one may shoot for years without meeting with any of them. They are the Green and Wood-Sandpipers, the Little and Temminck's Stints, the Dusky Redshank, and the Phalaropes.

Coincidently with this extensive "through-transit" in August and September, there also occurs the arrival of those hardier members of the same great bird-family which mean to make our shores their winter home. These are chiefly Curlews, Godwits, Knots, Redshanks, Grey Plovers, Turnstones, Dunlin, Purple Sandpipers, Sea-pyots, and, in a less degree, the Sanderling. The latter might almost be included in the former category, so very few are ever found here in winter, though abundant enough in August.

The months of August and September, it will thus be seen, are a period of great activity among the feathered tribes of the coast, and an interesting period to spend among them. Rambling about on a fine autumn day over the rolling wastes of sand, one can enjoy many charming views of bird-life. Suddenly one finds oneself almost in the midst of a flock of graceful little creatures—Dunlins, Ring-Dotterels, and Sanderlings, all mixed—which, among the myriad small pyramidal piles cast up by the sand-worms, had escaped observation at first. So tame are they that one can watch, close at hand, their pretty postures and agile movements as they dart about, nimble as mice, each little form reflected on the mirror-like surface of the wet sand. Further on, close up to the sand-links, are little parties of the Curlew-Sandpipers, and where a mussel-bed has created a mixed deposit of mud and sand, will be found the Turnstones and a few Grey Plovers. The great chattering flights of Godwits are always wild and cannot be approached on foot; they and the Knots mostly frequent the mud-flats. Overhead, behind, and on every side resounds the incessant scream of the Terns, busily fishing in the little pools left by the receding tide. Head-first into the shallow water they plunge, one after another, completely disappearing for a second, and hardly will the small fish escape their scissor-like bills. In sheer exuberance of spirits they scream and dive, and dive and scream again. Presently there is turmoil; one of their persecutors, the piratical Skuas, has come on the scene, and the plucky little birds at once unite in a combined attack on their common enemy.

Some very interesting problems centre round these little August migrants. Few problems nowadays remain unsolved —that is, few of what may be called questions of fact, though dozens of mere abstruse interest still remain unanswerable. Who, for example, can say why the Godwit, Knot, and Curlew-Sandpiper should turn wholly rufous-coloured in summer, while the allied Curlew, Whimbrel, Greenshank, and others remain unchanged? Why should the Plovers and the Dunlin in summer acquire black breasts, while the Redshank, Sanderling and several other *Tringæ* remain white? Such questions will probably ever remain unsolved, though, I suppose, by the Darwinian hypothesis, there should be some "first cause" assignable, however remote, chimerical, or shadowy it may be.

But of the few practical problems still unsolved, there remains the question: Where do the common Godwit, Knot, Sanderling and Curlew-Sandpiper breed? Whence come they in myriad hosts every August to our shores? That none of these breed in Europe seems tolerably certain. Spitzbergen and Novaya Zemlya certainly do not attract them, nor does Franz-Josef Land appear in the least degree likely to prove their *incunabulum.* The Grey Plover and the Little Stint were formerly in the same category; but the Siberian explorations of Dr. von Middendorff and other foreign naturalists, and, later, of our countrymen, Messrs. Seebohm and Harvie-Brown, have removed both these species from the list of "temporarily missing": though, of the four first-named species, little or nothing was seen. On the American side of the Atlantic, the Knot and the Sanderling have been discovered breeding on the North Georgian Islands, on Smith Sound, and elsewhere; but it is impossible that all the millions of these birds can be of Transatlantic origin. They *must* have some Old-World resort.

Kindly, "general reader" (if you have so far borne with me), give this matter a moment's consideration. It is *not* a question merely of where certain birds' eggs are to be found, but a mystery whose solution must interest any inquiring mind. Here we have four kinds of birds, all four immensely numerous as species, and two of them of tender and delicate constitutions, intolerant of any great degree of cold. They all pass to the northward of our islands in May and return in August or September. During this interval they have reared their young. Their natural economy demands during this period of absence a region where the climate is mild and warm—a spot where plant and insect-life abound. Their summer home can be no small district—their immense numbers preclude this. No limited area, no groups of rocky islets, or the like, would accommodate them at all. Wherever it is that they go, there *must be*, somewhere, a vast and comparatively fertile region which forms their summer home; but we know not where it lies. As before remarked, its discovery is not necessarily a mere question of abstract scientific interest, but might possibly (or even easily) prove of permanent value to trade, commerce, and the human race.

Where is the unknown region? The idea of a warm circumpolar sea (inaccessible to man through ice-barriers in lower latitudes) must now be regarded as exploded. The hypothetical submarine current of warm water, passing deep beneath the known ice-fields all round the Pole (which alone could produce this open sea), has been proved to have no existence. The unknown region must be nearer home than the Pole. At the moment, Siberia appears to be the only possible answer, for not even the energies of a Seebohm, or of a thousand Seebohms for a lifetime, could explore the whole of that vast region extending from Europe to Far Cathay. To the scientist and to the pioneer of commerce who have the time, the means, and, above all, the endurance, Arctic Siberia probably presents the richest field for the investigation of many unsolved problems and undeveloped resources.

In addition to enjoying this distinction—of being the ONLY British bird whose eggs have never yet been discovered by civilized man*—the Curlew-Sandpiper is also a remarkable example of wide geographical range, and of perfectly marvellous powers of flight in a species no bigger than a Snipe. Though breeding in the remote and unknown penetralia of the Arctic regions, yet in winter Europe is not wide enough

* The Knot and Sanderling have been found breeding, as above stated, in Arctic America, but *never*, in any single instance, on this side the Atlantic. The eggs of the common Bar-tailed Godwit have been discovered in two or three isolated cases in Finmark, Lapland, and, *perhaps*, in the Taimyr Peninsula. Such instances are, however, merely accidental and exceptional, and in no way indicate where the vast bulk of these birds breed, and the species is not found in America. The Curlew-Sandpiper has not yet been found nesting either in the Old or the New Worlds.

for it. During September they pass southwards. Some travel by the western route, *viâ* Norway, England, France and Spain; others cross Europe by the lines of its great rivers; while a third contingent, undaunted even by such barriers as the Central-Asian deserts and the twenty thousand feet of the Himalayas, boldly traverse the whole extent of the Asiatic continent at its widest points. Of the first two sections, a few winter in the Mediterranean; but the majority push forward along both coasts of Africa, and, crossing the tropics, winter in vast numbers on the shores of Cape Colony, Natal, and Madagascar. Then the Trans-Asiatic section, after reaching the coasts of India, Burmah, &c., continue their southern career through the Malay Archipelago, and eventually winter in Australia and New Zealand. To say they *winter* there is, of course, a misnomer, for it is obvious that birds which can thus transfer their home bi-annually from one hemisphere to the other, practically exclude that period from their chronology. The Curlew-Sandpiper enjoys the advantage of perennial summer. They, or at least the majority of them, pass what is our winter in the summer of the southern hemisphere. In (our) spring they begin to move north again. Even in Australia they are obtained in April in full summer plumage. Early in May they reappear on the Mediterranean. In mid-May I have shot them in Andaluçia, together with Grey Plovers, Knots and Whimbrels, in perfect breeding plumage. During that month they traverse Europe, and by June have again disappeared from our view in the mists of the unknown north.

One more example of the utterly inscrutable dispositions of Nature with regard to the migrations of this bird-group—a volume might be filled on the subject! The first arrivals on our coasts in autumn are composed (in several species) *exclusively of young birds*, then only a few weeks old. The parents not having completed their moult, are not ready to leave their northern home till a week or two later. Thus these infantile creatures, still partially downy and hardly complete in feather, are able, without knowledge, experience, or guide, without pilot or compass, safely to traverse thousands of miles of unknown space. Yet, generation after generation, they arrive with unerring regularity, punctual to a week—almost to a day—at the very spots, often the identical creeks, streams or marshes to which, for untold years, their progenitors have also steered their course. Some species take distinct routes—like the Cunard liners—for their northward and for the southward journeys; of others the young birds affect one route, while the old travel by another. To this point I will refer presently.

Turning to a purely sporting point of view, the larger waders, even in early autumn, are usually wild and watchful birds, and by no means easy of access, even to a punt, though

"ON THE SCAP-POINT."

a fair number can sometimes be thus obtained, chiefly from the smaller flights—the main bodies, at the same time, being often wholly unapproachable. A gunning-punt, moreover, despite her slight draft, is but ill-adapted for the pursuit of this class of sea-game, so extremely flat are their favourite resorts. There are two methods which are more effectual to secure them. At full spring-tides, when the sea comes right up to the sand-links, or main coast-line, the waders are driven up within shot of the banks or other cover. But it often happens that they will then betake themselves to the open fields or the refuge of the sand-bar, rather than incur such risks—at full-sea, I have seen large ploughed fields quite "grey" with Godwits. The other method is to lie in wait for their flights at ebb and flow, when they have usually more or less regular lines, a knowledge of which will afford as much shooting for a few minutes as a grouse-drive. They leave their feeding-grounds when the flowing tide covers the flats, returning as soon as the sands begin to dry again, at the ebb. The points they pass over in coming and going to their interim resting-places are the position for the gunner.

SOME SPECIFIC OBSERVATIONS ON THE WADING BIRDS.

Godwits.—These are birds whose plumage and distribution have very often been wrongly described. The letter-press appended to Bewick's inimitable woodcuts is wrong in almost every fact stated—both generically and specifically. The descriptions in Morris describe nothing at all; Montagu at least illustrates the outer darkness in which scientists of that era were groping their way; and I find it impossible to reconcile the opinions even of the latest authorities with my own observations on the north-east coast. Colonel Hawker professed no knowledge as a naturalist, but, at any rate, he wrote FACTS; his information on this, as on every other point, was *sterling*, so far as it went, and his remarks on Godwits (Ed. X., p. 226) are probably better worth reading than all the speculations of contemporary ornithologists.

There are two species of Godwits—the Common Bar-tailed Godwit and the Black-tailed. The latter, however, may at once be dismissed as all but unknown on the N.E. coast, save as an accidental straggler on migration, usually in September. It was this species which formerly bred in the "fens": nowadays there are no fens, consequently no Godwits.

The species referred to throughout these chapters is the Common Godwit (*Limosa lapponica* of Linnæus), generally called in books the "Bar-tailed" Godwit, though its tail is *not* barred, except in the young (*cf. infra*). The old Godwits, while here, have the tail plain ash-blue like the rest of their winter plumage. A few of the outside feathers, it is true, exhibit white splashes or indentations on their inner webs; but these are not *bars*, and cannot be seen except when the tail is widely spread open.

There is an extraordinary range in the colouration of this species. There are three distinct stages. The young in first plumage are speckled brown and buff. Adults are chestnut-red in summer; ash-blue in winter. The two first-named stages have the barred tail. Between these three extremes there is an infinity of intermediate grades—the more so as Godwits take two years to acquire their full adult dress; and, indeed, the plumage of the second year is so distinct that it should, perhaps, be described as a *fourth* stage. It has a character intermediate between the first plumage of the young and that of the adults in winter—having both the shaded ash-grey ground-colour of the latter, and also some of the speckled or mottled features of the former.

GODWIT, ADULT. (JANUARY.)

It is these endless variations which have perplexed and confused scientists for generations, and at the present day there is, perhaps, no one who fully understands them—none certainly better than my friend Mr. C. M. Adamson, whose study of the Godwit ("Scraps about Birds," pp. 40 and 158) should be digested by any one who wishes to unravel these intricate problems. Some ornithologists appear to be unaware of the blue (winter) stage, which, at least on this coast, is by far the commonest of all. This appears rather unaccountable in the case of so extremely abundant a winter bird, of which numbers of specimens might easily have been procured for examination. The explanation is perhaps to be found in the fact that modern authorities of the first rank state that the Godwit is *not a winter bird at all.* My friend Mr. Howard Saunders, in his excellent revision of Yarrell's "British Birds," describes it as a bird mainly of double passage (*i.e.*, spring and autumn) and rarely occurring on our coasts in winter: and Mr. Seebohm, in his "Monograph of the *Charadriidæ*," says that only "a few stragglers occasionally remain during winter." Whatever their distribution may be elsewhere, this is certainly not the case on the north-east coast, where the Godwit is one of the most abundant of our winter wildfowl, and may always be found, thousands strong, throughout the hardest winters and most protracted frosts.

These winter Godwits are all blue—some in the uniform ash-grey dress of maturity; the young in the shaded or marbled stage still showing traces of their former mottled plumage, especially on the tertiaries.

In their handsome ruddy summer plumage, the Godwits are all but unknown on this coast. On their southern passage a rare chance straggler in red dress may once in ten years be obtained among the thousands of young birds in August; but it is clear the old birds at that season take a different route to the young. They probably pass to the eastward, *viâ* the Baltic and Continental Europe.

On the northward passage (in May) adults appear to be entirely unknown. Our winter residents leave before acquiring summer dress, and the migratory Godwits from southern countries appear to strike off from the British coast about East Anglia. That is, they "take their departure" for the North Sea passage from, say, Norfolk, instead of continuing their northward course along the line of our coast.

KNOTS, arriving in August, are extremely abundant all the winter. A few old red birds (in summer plumage) are occasionally met with in August; but appear quite unknown in May, when going north.

GREY PLOVER arrive about mid-September, *all young birds*; and large numbers spend the winter here. Adults in breeding plumage evidently take a different course at that season, for they very rarely occur on this coast. It is, however, noteworthy that the very few which are obtained occur *earlier* than the young birds—namely, in August, nearly a month before the young arrive—the reverse of the general rule with the allied species.

Grey Plover never assemble in large packs, ten or a dozen being the maximum numbers usually seen together. Their disposition is social, rather than gregarious as a species, and almost every great cloud of Dunlins, &c., has the company of two or three Grey Plovers. I have noticed them associating with almost all the different waders, and even with half a dozen Ring-Dotterels on the rocks.

SEA-PYOTS.

They are easily distinguishable by their note from the Golden Plovers ; but, indeed, the latter are very seldom met with in the "slakes" in winter, except in severe weather, when driven off the moors and inland haunts by snow and frost. The Grey Plovers, on the other hand, never leave the tidal area.

Although these Plovers and the Knots appear quite impervious to the utmost degree of cold experienced in our British winters, yet large numbers of them travel a great distance to the southward of our islands at that season. They are common in the Mediterranean, and in the south of Spain

I have witnessed the passage of both species (from Africa) in mid-May, in full breeding plumage.

RUFF and GREENSHANK: only obtained in August and September.

SANDERLING, abundant at same season, but a few occur throughout the winter, though sparsely. Adults in breeding plumage are occasionally obtained on northern passage in spring.

OYSTER-CATCHER.—On January 20, 1882, observed one of these birds swimming about in deep water. Expecting it was a "pensioner," or wounded goose (which always separate themselves from the packs, and are found singly thus), we punted out to it. To our surprise, the bird proved to be an Oyster-catcher, and after rising close to us, flew round, and again settled in deep water. Except on this occasion, I never saw a wader swim of its own accord, though, *when winged*, the Oyster-catcher can both swim and dive fairly well.

GREBES AND DIVERS.

THE Grebes are a class of fowl one frequently meets with both in the sheltered waters of the harbours and on the open sea. Four species are met with on this coast, of which the most abundant is the Sclavonian Grebe, the rarest the Eared kind. The only Eared Grebe we have ever obtained was shot by my brother J., on February 6, 1879, while running for shelter into Holy Island harbour. The above bird weighed 11¼ oz., a number of Sclavonians ranging from 11½ to 13 oz. These two species closely resemble each other in size and general appearance (in winter), but are distinguishable by the up-turned bill of the Eared kind, and by its white neck, that of the Sclavonian being dusky.

The Great Crested Grebe is not infrequently seen: an adult female, shot February 1881, weighed 31 oz., and its gullet contained several small fish. The Red-necked Grebe also occurs, but less frequently than the last-named. The Little Grebe never appears ón salt water, though they frequent the backwaters of Tweed and other Border rivers during winter.

DIVERS.—Of the three species the Red-throat is so well-known a coast bird as to require no remark further than that I obtained a specially fine example in full summer plumage, with red throat, on March 26, 1881. This bird I gave to Mr. Hancock.

The Black-throated Diver is by far the most scarce. I have only met with it on two or three occasions, and shot one at sea, January 22, 1880. It was a male, weight 5 lbs., length 27 inches, expanse 42½ inches. In the winter plumage, with its marbled blue and grey back, this species appears to approximate most closely to the Northern Diver; but, unlike that species, it is a strong and good flier, always ready to take wing on the approach of a boat, which I have never seen the Northern Diver do. The latter of course can, and does fly well, during the breeding season, and when on migration. I have seen a pair, in March, passing over northwards very high, and at great speed.

The Great Northern Diver is not infrequently met with on the Northumberland coast, a few occurring almost every winter. The following are the dimensions of an adult pair shot by myself in January :—

	Weight.		Expanse.		Length.
Male	12¾ lbs.	...	54 inches	...	36 inches
Female ...	8 „	...	48 „	...	30 „

CORMORANTS.

The male (shot January 19th) still retained a considerable proportion of the handsome spotted plumage of the previous summer on the wings, but the back and neck were in full winter dress. His gullet contained two or three good-sized flounders. The beak of this species *in winter* is pale horn, or ivory colour.

Both Grebes and Divers are remarkably variable as to the numbers which come in different years. Some seasons both kinds are fairly numerous; in others one may hardly see half a dozen all through the winter. They are found, both inside the estuaries, where they fish in the sandy "guts," and perhaps more plentifully on the open sea. Grebes often associate into little flocks, but the Divers are usually solitary.

CORMORANT.—A pair of these birds, shot right and left at day-break one winter's morning by Mr. Crawhall, weighed together no less than 17 lbs.—a surprising mass of bird-flesh to bring down with a double shot of a 12-bore! The larger bird, a 'champion' Cormorant, turned the scale at 10 lbs.

WILDFOWLING IN MILD WEATHER.

"BLANK DAYS" IN JANUARY, 1886.

Most descriptions of wildfowling which have been published relate to successful attempts made under the favourable auspices of severe weather. One seldom hears anything on the subject, or of what has been done (rather, in many cases, *not* done) under the reverse conditions. Perhaps, therefore, the following account of (as regards spoil) a somewhat fruitless expedition may be of interest, both as showing the other side of the picture, and illustrative of the habits of wildfowl during mild seasons.

Possibly many might object to the winter of 1886 being described as "mild"; and in point of fact it was the most irregular winter in its intensity and erratic in its distribution, that has occurred in the writer's experience. In the same county, within twenty miles of each other, we had both winter and spring simultaneously: where I write (Feb. 8) there is not a sign of winter, but an hour's journey or so inland the snow lies several feet deep, roads and railways are blocked, and all the rigours of a most severe winter prevail. Throughout the north of England generally the snowfall has been local and "patchy," in some districts the frost having held almost continuously for weeks, while elsewhere not a vestige of snow was visible. Thus on the inland moors, for example, there have been successions of heavy snowfalls, alternating with partial thaws and renewal of frost, this causing the half-melted snow to freeze again into a crust so hard that neither sheep nor Grouse can reach their food. The hill-farmers are at their wits' end, and the local papers almost daily contain reports of the occurrence of Grouse and other moor-birds in most unusual localities, very many miles from their accustomed haunts. In this county (Durham) Grouse have straggled down to within three or four miles of the coast—the nearest moors being thirty or thirty-five miles inland—and for "the first time on record" we have had them in our neighbourhood at Silksworth, only some three miles from the sea. Such phenomena are similar to what I recollect in the severe winter of 1880-1 and some former ones; but hitherto only in winters of universal severity, and are remarkable in a season of such sporadic intensity as that under notice.

On the coast, on the other hand, there has been little or no severe weather; what frost there has been seldom lasting more than a few days, insufficient either to bring over fowl in any quantities, or to "tame" them when here. Towards the end of January I paid a three days' visit to an estuary which is usually frequented at this season by fair quantities of ducks and Brent Geese. The numbers of the latter this year I estimated at only one-twentieth of what we have in hard winters, and considerably less than I recollect seeing in any former year. I must guard myself against appearing to infer that the state of the winter here is the sole factor in influencing the quantities of wildfowl which migrate hither at this season. Brent Geese especially are so strongly hyperborean in their affections, that their movements are regulated almost exclusively by the state of the winter and extent of ice in Northern Europe, and but little, if at all, by our local conditions, as was demonstrated a few weeks later (see next chap.). Inferentially, the winter must have been unusually open farther north, though I have no direct means of knowing. Mallard and Wigeon, however, being fairly plentiful, appeared to offer the best chance of scoring, so we went afloat on about a quarter flood at five in the morning, in hopes of getting a good shot at the ducks before they went to sea at day-break. The night at that hour was bright and calm, starlight and a third-quarter moon affording quite sufficient light for a shot to the westward. We had some miles to "pole" before reaching the favourite feeding-grounds of the ducks on the *Zostera*-covered mud-flats; and, ere we reached our destination, a most unfortunate change came over the night, which completely altered our prospects and reduced the chance of success to zero. About half-past five we noticed a slight haze beginning to rise off the water, which gradually and rapidly increased in density. Spreading over the face of the heavens, in half an hour it obscured moon and stars, and enveloped us in hopeless darkness. We were now in the best of the feeding-grounds, and amidst the gloom we heard around us at intervals the enticing "talk" of both Mallard and Wigeon, and anon the strange hoarse bark of the Sheld-Duck. But not a fowl could we see. We groped about helplessly, hopelessly, amidst that Cimmerian darkness, while the prizes of our hopes were plentifully in evidence around.

About the period when (by the almanac) daylight should have appeared, we heard the clanging chorus of the Brent Geese arriving in from the sea, and presently we made out a small "bunch" of a score or so—an indistinct line of grey dots on the grey water—some two hundred yards from us. Geese are usually dullish fowl *in the dark*; so we cautiously "set" to these with renewed hopes. Dusky as it was, however, the geese were fully on the alert, for they rose almost immediately, and, though I risked a long shot with the stanchion-gun, it was not responded to. We observed, however, that a single goose lagged considerably behind. This shot was rather disappointing, we having counted on securing a couple or more. The elevation and instant of firing were, I knew, both correct; but in the fog the geese loomed large, and were probably farther off than we had calculated.

The density of the fog relaxed a little as the sun rose, and we proceeded on a cruise round the whole of the mud, with the result of ascertaining that all the ducks (except a few Sheld-Duck) had gone to sea at dawn. There only then remained our friends the geese, to which we directed our attention. Several times we advanced on their main line, but with one unvarying result. Whether we paddled, "poled," or sailed, the watchful fowl rose at (roughly speaking) six hundred to eight hundred yards—a distance at which one might suppose a gunning-punt, end on, was almost invisible. Each time we noticed our "pricked" goose lagging behind, though he always managed to rejoin his company. The Sheld-Ducks, too, proved quite unnegotiable, as is usually the case in mild weather. Several small lots were busily feeding on the mussel-scaps, often in company with such very unsuspicious fowl as Oyster-catchers; but the Ducks always had one sentry, bolt upright, and ere the punt glided within two hundred yards his broad goose-like pinions were spread, and silently and without a sign of warning they left their more simple friends behind.

By midday the tide was half-ebb, and the mud-banks were reappearing. Simultaneously the wading birds, in their varied kinds, began to congregate from all sides. I think I have never before seen, in many years' experience of wildfowling, such immense quantities as we had of these birds that winter. I hesitate to attempt to estimate numbers, but may mention that a single flock (or rather a cloud) composed chiefly of Godwits and Knots, certainly extended to a quarter of a mile in length, and appeared to be about twenty-five or thirty birds abreast. Its numbers can be roughly computed by my readers. In addition to these, perfect hosts of Dunlin and small waders covered the mud as it dried, and the volume of tiny voices came rolling across the waste in a sea of undulating sounds. The rest of the day was, for lack of better game, devoted to the Godwits, Knots, and Grey Plover, but with only meagre results. On the wide-stretched, dead-level flats of mud and sand it seldom happens that the bulk of the waders are congregated within shot of water sufficiently deep to float a punt. Indeed, even the lightest gunning-punt is, in such places as I refer to, but ill-adapted for killing any great quantities of this sort of sea-game, and I have never yet succeeded in making a really satisfactory shot at Godwits in winter with the punt-gun, nor heard of any one else doing so. At low water in the evening we took a cruise round the deep water channels or "guts" to look for divers, which at this time of the tide are confined to these channels. But these birds were conspicuous only by their absence. We found nothing but a few Mergansers and a single Golden-eye—both these always inaccessible to a punt. Not another Diving-duck, or even a Grebe, was to be seen, and of the Colymbi, a couple of Red-throats and one Great Northern Diver were all I observed during the three days. The latter was on the open coast, unconcernedly feeding amidst a boiling surf that was breaking outside a reef of basaltic rocks. On seeing us he dived, reappearing a good quarter mile to seaward.

The next day was again a blank. A whole gale from the south-east made any idea of going afloat quite out of the question. At the morning flight I did manage to drop a Mallard at the harbour entrance, but even this solitary spoil I was not destined to get, for, with the predatory instinct of his tribe, a fisherman-gunner, who had taken up a position behind me with his dog, quietly retrieved the duck as it drifted ashore, and decamped—a bit of by-play I did not observe at the time, it being still dark. Eventually I secured a Heron and a couple of Curlews in return for two bitterly cold hours' lying out on the weed-covered rocks. In spite of the heavy sea that was running outside, the great bulk of the geese left the harbour, as usual, at dusk. They went to sea in a single body, and at least one hundred yards high, though the gale blew dead against them. Only one small lot of about two score remained inside sitting in the "deep," where no punt could approach them.

An hour before daybreak on my third and last day again found us in our former position at the edge of the mud-flats. But once more Nature persisted in frowning on our endeavours. Our first attempt she had frustrated by a fog; with a storm our second. No such adventitious phenomena were needed to be invoked by her to thwart our hopes this morning. The obstacle now presented was simply the daily variation in the tide. Before us, on the mud, we made out two big packs of Wigeon, all unsuspectingly feeding under the feeble rays of the waning moon. In less than an hour they would, under the friendly shelter of darkness, have been in our power. Had we now but the conditions of tide which prevailed yesterday, we should have been nearly sure of a heavy shot; but to-day we knew full well that, before those sixty minutes had run their course, the treacherous daylight would have appeared, and—even if the Wigeon remained inside—revealed to the whistling phalanx its threatened danger. Still, after the storm of yesterday, we cherished a hope that the Wigeon would hesitate to take the sea; and as the daylight broke we had the momentary satisfaction of observing that one of the two packs *did* elect to remain on the smooth and sheltered waters of the estuary. After many gyrations, and a loud chorus of their musical "Whee-yoo!" this company settled on the deep water some half a mile outside our position, the other section going out to sea. But Fortune was only toying with us, for, after patiently waiting till the flowing tide had carried them into shoal water where an attack was possible, we had once more to submit to failure. This pack was composed exclusively of Wigeon (the other having been at least half Mallards, which are always easier of access by daylight), and refused to be cajoled, rising some three hundred yards off, and following their companions to the open sea.

Up to this point our lack of success had arisen exclusively from circumstances entirely beyond our control—fogs, storms, tides, and the like. But now we *did* throw away a last remaining chance to some extent through an error in judgment. Far away along the edge of the rapidly disappearing mud we descried our "pricked" goose—a black dot bobbing about on the tide; not another fowl was in sight. A careful scrutiny of the remaining banks with the binoculars satisfied us that no ducks remained anywhere near us, so I proceeded to gather the "pensioner." Alas! as the report of the cripple-stopper rang out across the waters, there rose from behind the bank of a tiny creek over a score of ducks—all Mallards—which, had we but been aware of their presence, would in all probability have fallen an easy prey. Fortune and the elements appeared to conspire to deprive us of the few chances which are the utmost that can be expected in a mild season, and with despondent hearts we watched the little string of Mallards, our last hope, speeding away to the open sea.

The rest of the day was spent in fruitless attempts to out-manœuvre the impracticable geese. Once only did we appear within "measurable distance" of getting a chance. This was by running down on them under sail, and luffing sharply as they crossed our bows to windward. But the breeze failed us just at the critical moment, and though I "tipped" the big gun and risked a very long shot, we got nothing.

And now I fear my reader's patience will be well-nigh exhausted with this long catalogue of misfortunes and mis-chances. Such, however, are from time to time the inevitable concomitants of coast-gunning *during mild weather*, and he will perhaps charitably remember that the experiences which will cost him but ten minutes to read, occupied the writer the greater part of three long days and nights.

WILDFOWL AND THE WEATHER IN MARCH, 1886.

The month of March presented such unexpected and remarkable climatic phenomena, attended by an almost unprecedented influx of wildfowl, that the following account of it may be an interesting record.

Up to the end of February the winter had been unusually irregular and local in its severity, and the quantity of wildfowl on our coast considerably less than had been the case for several years. But on the morning of March 1st, we awoke to find a heavy and persistent snowstorm, driving before a south-easterly gale. All that day and the following night the storm and gale continued without intermission. The snow was of that fine dry powdery description which forms the most dangerous drifts, and the morning of the 2nd found us, in the north of England, cut off from communication with the outer world: we had no post, no newspapers, the snow lay piled in huge drifts, and—still worse—there were no signs of abatement. The wind veered to the north-east, but the snowfall continued all that day and night. After nearly forty hours' incessant snowfall, the morning of March 3rd at last broke fine, though the wind still blew strongly. During this time we had been to a great extent deprived of news from outside, each town and village being cut off from its neighbours, and it was only as communication was gradually restored that we learnt the full extent of the storm.

As soon as railway communication was partially reopened, I received a letter from my puntsman (on March 6th), telling me of the arrival of the Geese. On the afternoon of March 2nd, he wrote, after some thirty-six hours' incessant snow-blast, the Geese began to appear in thousands. Flight after flight, all that afternoon, they came pouring in from the sea; their dark columns all blended with the driving snow, and alighting in dense masses in the harbour, and even along the mud, close to the village. During the night the arrival still continued, as could be judged by their notes, and on the morning of the 3rd the numbers which had come were roughly estimated at fifteen thousand to twenty thousand. Two Swans also passed to the northward that day, the first seen this season, and fresh bodies of Geese kept coming in all day from sea, until the total aggregate could not be estimated (as I saw myself) at less than thirty thousand; and this in a single harbour, where there had not been over four hundred or five hundred Geese all the winter.

Now in connection with this extraordinary influx of Geese on March 2nd and following days, it is interesting to know the state of the weather abroad, and especially in Denmark and the Lower Baltic, whence they had most probably come. The frost in Denmark, which had been extremely severe towards the end of February—a friend who left Copenhagen on the 28th luckily got across in the last steamer which could leave Korsoer for Kiel—was intensified in March; and on the 2nd (the very day of the arrival of the Geese here), the following telegram was despatched by Lloyd's agent at Copenhagen: "The frost continues with increased strength, and the navigation is almost entirely stopped between the Scaw and this port. Powerful steamers have succeeded in forcing their passage; but seven or eight steamers are reported to be fast in the ice." Later telegrams reported that "the Cattegatt is full of ice, and navigation most dangerous. All the Baltic and Danish ports are closed, and the frost still continues." In short, the whole of the sounds and harbours of the northern coast of the Continent were frozen up in the early days of March, and the "grand army" of Geese which usually winter there, at once crossed over to seek refuge on our side of the North Sea.

On March 8th, the North line having been dug out, and several embedded trains released from the depths of the snow drifts, I went down to an estuary in North Britain, and was afloat next morning by daybreak. It was a bitterly cold day, with 16° of frost, and a cutting wind off the snow-clad hills. The rounded decks of the punt were soon encased in a sheet of ice, and the sea-water froze into icicles along the barrel of the punt-gun and elevator. But, cruel and biting as was the cold, the marvellous spectacles of bird-life we witnessed that day were ample compensation. Never have I seen such sights as were presented by the multitudes of Brent Geese. Words fail to convey any adequate idea of their numbers, and of the effects produced by the disciplined evolutions of their vast hosts as they wheeled and manœuvred in the air. In roughly estimating their numbers at something like 30,000—more than double the number we had in the severe winters of 1878-9 and 1880-1—I fear I may be suspected of exaggeration. But these numbers are probably not very far wrong.

In spite, however, of the numbers by which we were surrounded, we found it no easy matter to get near them, or to lay hands on so much as a single goose. Hour after hour was spent in fruitless efforts, which, up to midday, were not rewarded by a shot. The tide was ebb, consequently the main bodies of the Geese were congregated upon the dry mud, far beyond our reach. There was, however, no lack of them in the deep water channels, or "guts," to which detachments of several thousands were continually resorting during the intervals of feeding, and where they sat in the water, splashing, washing, and preening themselves. These, however, proved so extremely watchful and wide-awake that, though I "set" to them at least a dozen times, I never succeeded in getting within fair shot. Two or three times I had small straggling "bunches," or the fringe, so to speak, of their main line within range; but with the enormous numbers in view, I was all anxiety to make a *heavy* shot, and declined to accept such paltry chances.

As the day wore on and afternoon arrived, with the forepeak still empty, we began to despair of getting to close quarters, and I risked a long shot, which secured five. After this it was a long time before I could get on any sort of terms with them; but as the flowing tide gradually covered the mud-flats, I began to get in touch of the main bodies of

Geese which had been feeding on the dry. Presently I approached the flank of one of these hosts of Geese which were feeding along the mud-edge. Before going into action I landed to reconnoitre the enemies' position, and shall not soon forget the sight I witnessed from behind a high bluff on the sand dunes. Commencing at five or six hundred yards from our position, the whole shore-line was literally blackened with masses of Geese, extending for some two miles along the shore. In places the line was thicker, in others more open, but nowhere could a break be seen, and the aggregation of bird-life formed a spectacle such as few have ever viewed. I was now almost at the end of the *Zostera* banks, consequently the Geese had to fly back; so I shoved out to try to intercept them, and as the rearmost files of the two-miles-of-geese crossed the bows, though at a considerable distance, I tipped the punt-gun and knocked down four, two of which were lost in the rough water outside.

So far I had done practically nothing; but luck was yet to come. Towards evening, after "poling" several miles along the shore, I again came up with one of those huge companies of Geese which I have tried to describe. Just as the sun was setting I commenced to "set" to them. The tide being now full flood, enabled me to advance on them from under the shelter of the snow-clad banks—a great advantage. They were, nevertheless, pretty wide-awake, and as the punt glided to some eighty yards' distance the whole mass rose simultaneously with a roar like thunder. At that range, however, the punt-gun cut a pretty lane through their black ranks. In all twenty-one Geese fell direct to the shot—a capital performance, at the range, for a small gun of under 60lb. weight, and throwing only 10oz. of shot. The cripples formed a tolerably solid flotilla at first, and the play of the cripple-stopper soon stretched all but one or two of the most lively on the sea; but it takes a long time to catch so many, and, with the increasing darkness, it was impossible to secure the latter that night.

One of the most delightful features of punt-gunning in very severe frost, such as then prevailed, is the opportunity of observing, at very close quarters, many birds which are ordinarily unapproachably wild. The intensity of the frost, covering the oozes with thin sheets of ice between tides, has the effect of making many fowl quite tame. Mallards especially were frequently passed within twenty or thirty yards, sitting asleep on the mud, with their bills tucked under their back feathers, others paddling about the water's edge, dabbling about among the sea-grass, all quite unconscious of our close proximity. The Mallards were in small bunches of three or four up to a dozen, and all these were the heavy native-bred ducks, driven down to the open water of the coast by the severe weather, their ordinary haunts on the moorland lochs being frozen and snowed up. These heavy ducks it is most unusual to meet with on the coast at this season, except under such exceptional climatic conditions as prevailed that March. They are easily distinguished from the foreign ducks by their extra size and tameness; moreover, the foreigners are not usually found in harbour by day. They, together with the Wigeon, to the number of perhaps a couple of thousand, spend the day in their accustomed resort, a secluded bay a mile or two along the coast, where they are safe enough from man and all his devices, and do not approach their feeding grounds till well after dark. Redshanks were frequently feeding within ten yards, up to their breast-feathers in water, and a pretty sight it was to watch the impetuous manner they tossed aside the floating weed to find some food which it concealed. Oyster-catchers and other waders were also extremely tame, and for the first time since January 1881, I noticed great numbers of Golden Plovers out on the salt-slakes. On one occasion I approached a long thin line of Knots on the mud-edge, all asleep, no heads in sight, and looking for all the world like a strip of rounded blue pebbles, except that they were raised a couple of inches off the mud. As these birds do not breed in England (or in Europe either for that matter), and neither they nor the Geese are included in the Wildfowl Act, I sent a charge from the cripple-gun athwart their line and secured eight of them.

On the evening of the last day a catastrophe occurred. It

was full tide, and just before dusk an immense body of Geese were feeding in to the shore. I crept close along the banks under the ice-edge; the Geese drove in with the tide, and in a few minutes I was, so to speak, in the midst of them. The water was of the deepest blue, and, in the bright rays of the setting sun, it fairly shone with the innumerable glossy black necks and snow-white sterns. But the big gun missed fire. Had she "gone," I must have killed Geese from forty yards up to one hundred. So near was I, I had time to pull out the double-10 from under the fore-deck, and stop a couple of those which had risen within twenty yards of the punt's beam.

I will now pass on to the departure of the Geese from our coasts. The facts in connection with their withdrawal are quite as interesting to naturalists as those which attended their appearance. For, just as their arrival here has been shown to have coincided with the closing of the North European sounds and harbours, so their departure was precisely, to a day, contemporaneous with the breaking up of the ice in those waters. First, as to the weather: after a partial renewal of the storm about the middle of March, a thaw became general here about the 18th. So rapid was the transition, that the snow in my garden, which lay between two and three feet deep on March 18, had entirely disappeared by the 20th. On the 23rd a brilliant crop of crocuses appeared; two days later the grass turned green, and all was summer that a short week before had been Arctic. The temperature rose from 16° on the 9th to 60° on the 24th! On the Continent the thaw was a few days later, as the following extracts from the daily papers show. On March 24 (thermometer here 60° in shade) a telegram from Bremen stated, "Ice breaking up in Weser and at Vegesack," and on the same date the port of Gottenburg was announced to be open, and the first steamer forced her passage through the ice from Reval. On March 25, a telegram from Reval reported, "The Baltic ports are now open for navigation; ice breaking up slowly." A Stettin report of March 26 states, "Complete thaw here: ice disappearing rapidly."

Several other telegrams also confirmed the above dates of the re-opening of the waters.

The departure of the Geese coincided precisely with these dates. Large flights left our coast on March 23, and still greater numbers on the 24th, many others doubtless leaving during the night. On the 25th and 26th the remainder almost all took their departure, and of the tens of thousands which arrived here on March 2, hardly two score remained on April 1. The Geese when last seen were steering due east, and very high, mounting higher in the air as they went.

The Wigeon disappeared during the latter part of the month, and by the 25th were nearly all gone. Swans occurred three times in March. In addition to those already mentioned as passing on the 3rd, five others arrived on the 27th, and remained several days, and on the 31st six more passed to the northwards. These three occurrences are all that have been seen on this part of the coast during the winter. It will thus be observed that, after one of the worst fowling winters on record, the most abundant sport was obtainable (with exclusively foreign fowl) in March, at the very season when some non-practical theorists would have us believe that all wildfowl require protection.

Later in the year, by a curious coincidence, I had another opportunity of observing the movements of the Brent Geese. I left England for Norway on May 25th, and early on the morning of the 27th, the Norway coast in sight, distant fifteen miles, we saw far astern, an immense body of Geese on the wing, looking at the distance like a small cloud over the sea. They rapidly overhauled us, though our steamer was making eleven knots under steam and canvas, and passed outside her, heading due North. There were many thousands of them in long straggling skeins, and at the speed they were travelling (say thirty or forty knots) would reach Spitzbergen in about forty-eight hours—that is on one of the last days of May—exactly the date when they are due there!

The two months which had elapsed since leaving our British coasts on March 26th, the Geese had evidently spent in North-Continental waters or Danish sounds.

SUNDRY INCIDENTS OF FOWL AND FOWLING.

THE foregoing chapters describe the ordinary daily life and habits of our coast wildfowl; but denizens of such bleak and exposed haunts are necessarily subject to all the vicissitudes of our winter weather, which often vary their daily routine. Thus, during very rough seas, wildfowl are unable to "weather it" outside, and are driven to seek shelter elsewhere. At such times the estuaries may sometimes be seen fairly packed with Wigeon, &c., at midday; but one can only watch them covetously through the binocular, for no punt can stand the sea.

The effect of sudden gales are interesting as showing that these are sometimes quite unforeseen by the fowl—usually fairly accurate weather-prophets. This the following couple of extracts from old shooting note-books, will serve to illustrate:—"January 5th.—Many hundreds of ducks left for sea early this morning; but about 9 A.M., a sudden easterly squall coming away and knocking up a nasty sea, the whole of them returned inside for shelter. Shot one Wigeon drake as they passed up, but there was too much sea on to follow them." The other incident was with the Geese, and occurred during very severe weather. It was flood tide shortly before dusk, and the Geese had just gone to sea in a solid body several thousand strong, when a sudden and severe gale came away from S.E., driving us for safety into the southern (and most distant) corner of the harbour. Here we had hauled the punt ashore and were trying to keep ourselves warm by running up and down the narrow interval between sea and snow—literally "between the devil and the deep sea!"—and with no very cheery feelings as to our prospects for the night in that desolate spot amidst snow-covered sand-hills. Just at dusk, however, to our great relief, we observed a coble coming across to us. To her we transhipped the gun, gear, fowl, &c., and taking the punt in tow, commenced, with all reefs in, to beat back to windward. Suddenly the whole army of Geese re-appeared—driven back from sea to the shelter of the harbour. It was a grand chance to "Up helm, and run in under them!" but prudence forbade. The punt astern, light as a cork, has no steerage when in tow, and had we executed any such sudden manœuvre, would have filled, broken adrift, and been lost. I managed to stop a couple with a large gun as they crossed our bows, but with the rising sea and deepening darkness, we failed to secure them.

To the naturalist-gunner, perhaps the most interesting periods are the commencement, and then the latter end of the season—say the months of October and March. In autumn, birds newly arrived from the barren uninhabited north are naturally less difficult of access than later on, when they have taken in and digested the whole system of fowling. This rule is not absolute—no rules are—still there are perceptibly greater odds on the gun in autumn than during the mid-winter months. Then, as winter begins to merge into spring, the fowler may again hope to secure a favourable chance or two, if he only has the luck to fall in with some of the passing bands of fowl which at that season are gradually moving northwards. I once witnessed a considerable arrival of Geese (on 1st March) from the southward. It was shortly after daybreak, and the birds were evidently tired with a long flight, for, though at the harbour entrance they received several shots from the "skirmishers of the coast," yet they all settled, some 500, within a mile or so, in a narrow "gut" where I was lying in the punt. Naturally expecting to find the Geese very wide-awake after the reception they had just received, I elevated the gun high for a flying shot, but, to our great surprise, got close up without any signs of alarm being observed. Their utter carelessness, however, actually proved their salvation and, from my point of view, a serious catastrophe. Being at close quarters, I decided to take the sitting shot, and drew in the elevator, taking point-blank aim. But I forgot to allow for the high trajectory at so short a distance—perhaps 40 yards, never having been so near to Geese before—and thus, although the gun lay aligned on the thick of a forest of necks, the whole charge passed *clean over them*, without touching a feather ! Later in the day I had a modified revenge. The same Geese were rounding a point of land when we "set" to them from inshore. Never were Geese so tame ! So near were we as they crossed our bows, that we could distinctly see their black paddles working away under their barred flanks. They were at the moment too straggled to offer a fair shot; but as they weathered the point and formed up inside, we had their company in flank and managed to rake them.

Similarly, on February 28th (1880), my brother W. was out in a punt at Teesmouth when a flight of 60 or 70 Wigeon arrived in from sea, and, after a few gyrations, pitched along the edge of a sand-bank. Here they allowed so near an approach that, though he had only a shoulder-gun, the 3 oz. of shot stopped no less than sixteen, of which fourteen were fairly bagged. The birds were, no doubt, beautifully lined out, still it was an exceptional shot for a shoulder-gun, and in broad daylight. Neither of these two incidents would have occurred except with passage-birds, wearied and resting after long flights.

Birds of prey can hardly be included in the category of wildfowl, and indeed a list of the Raptores which are met with on salt water would be very nearly as laconic as the well-known work on the "Reptiles of Ireland"—there are no birds of prey on salt water. Yet the two following instances of the occurrence of these rare visitors are perhaps worth recording in this chapter of odds and ends. In the very hard weather towards the end of January, 1881, a large Eagle (probably the White-tailed Sea Eagle, *Haliaëtus albicilla*) appeared on the Northumbrian coast, near Goswick, and remained for several weeks, frequenting the slakes, where he fed on the plentiful supply of "pensioners" which that hard winter produced. This Eagle was seen daily by fishermen and others, and of course strenuous efforts were made to secure him, but always in vain, though several guns were often engaged at a time. He appeared to roost among the sand-links, which are very extensive, at times alighting on some disused buildings, and by day was usually seen sitting on the sands. I did not chance to fall in with the Eagle on any of my visits to the coast, but Mr. Bell, the mail-carrier, tells me he saw him almost every day, usually "sitting on the sands eating a 'Ware-goose' (= Brent), and with half a dozen Grey Crows perched all around and close to him." He remained till well on in March, when, of course, the supply of "pensioners" ceased to exist.

On December 8th, 1885, about thirty Geese arrived, and my boatman launched the punt to go in pursuit. He noticed they were unusually tame, especially as several Wigeon were in company with them. He had already got within easy shot, when suddenly a "Glead" appeared, having apparently descended from the skies, and hovered over the Geese within a yard of their heads. Three times they sprang, but, not daring to fly, splashed straight down again into the water just where they had risen. S. then fired, killing three Geese and two Wigeon, and the 'Glead' made off for the main land. From the description given me, this Glead was no doubt a Buzzard, being a large dark-brown bird with broad heavy wings and straight-edged tail. In answer to a question as to size, S. remarked that it "was not half so big as the Eagle (of 1881): that bird was as large as a canny-sized laddie sitting on the sands!"

In cruising year after year over the same grounds, one notices small geological changes (if the word is admissible) constantly going on, and the causes are often interesting. A little mussel-spawn happens accidentally to become deposited by the current on *sand*, instead of on its natural bed of mud. As the young mussels grow, the mud also appears to grow around them—the mussels create it. After a while the *Zostera* begins to take root, and in a few years a stretch of several acres of mud, with all its peculiar vegetable and crustacean productions, has from such small beginnings been created, in the midst of sand. This in time forms a fresh haunt and feeding ground for wildfowl. Ducks are now regularly found where none were ever known before those vagrant particles of mussel-spawn drifted on to the uncongenial sand. Of course such increases are counterbalanced by denudations elsewhere—the tide here and there sweeping away the superstratum of mud, and laying bare the non-productive sand below. Sand is proverbially a shifty substance, and every year its local geography alters more or less; the old channels disappear, and new ones open up through the midst of what was before quite solid ground.

Wildfowling is essentially a "waiting game," and the long hours one often has to spend in the creeks "waiting on" for the tide, or for the fowl to appear, give one plenty of opportunity of becoming tolerably well acquainted with, at least, the superficial features of the bleak spots around. Just examine the space that lies within arm's length of the boat—what a wonderful microcosm exists in every yard of ooze! The profusion of marine life is bewildering, and a multifarious struggle for existence rages as keenly here, out on the deserted oozes and sand-flats, as on the thronging Stock Exchange or the precincts of Lombard Street. What are those little coteries of Dunlins finding on the bare sand—what is it that impells them to dart hither and thither, as alert and active as frightened mice? Apparently there is sand, nothing but sand. But to them that sand is a perfect mine of wealth, in the form of countless tiny insects, crustaceans, sand-worms, and the spawn of an infinity of minute forms of life. The mud too has its swarming population. Those holes which everywhere perforate the banks of its creeks are the homes of the clam; its surface is traced in every direction with the trail of the wandering periwinkle—the objective of the "popular pin." These in warm weather lie scattered broadcast; the mussels cluster in dense groups. But why are all these empty shells strewn around? Examine one against the light, and there will be observed a tiny circular hole in its side. This is the work of the dog-whelk, a very cannibal among molluscs. With his strong-toothed proboscis, he drills a hole right through the hard calcareous armour of poor *mytillus*, and proceeds to feast on his succulent interior. But the whelk, too, has his enemies; for, on picking up half a dozen shells, several are found to be tenanted by that strange usurper the hermit-crab. . . . . But our investigations are suddenly cut short. "Look out, sir!" shouts our puntsman; "here come the Geese!" and in one moment we are aboard, and lying as flat as any oyster.

The following list of the local names by which the different wildfowl are known on parts of the north-east coast may perhaps be interesting to the etymologist:—

Grey Geese (generally) . .	"Grey-lags."
Brent Geese	"Ware Geese."
Mallard	"Mallart."
Wigeon	"Whews."
Golden-eyes	"Wigeon."
Scaup Duck	"Covie."
Longtailed Duck . . .	"Jacky Forster."
Eider Duck	"Culver."
Scoter (Black Duck) . .	"Sea-hens."
Merganser	"Yarrell."
Northern Diver . . .	"Nauk."
Red-throat Diver . . .	"Lion."
Grebes	"Tommy Allens."
Coots	"Belpoots."
Guillemot	"Willock."
Black Guillemot . . .	"Sea-pigeon."
Cormorant	"Gormer."
Whimbrel	"Curlew-Jack."
Godwit.	"Speethe."
Oyster-catcher . . .	"Sea-pyot."
Turnstone	"Brackett."
Purple Sandpiper (&c.) . .	"Tinkers."

The pronunciation of the word *Yarrell*, when subjected to the Northumbrian "burr," is well illustrated by the manner in which my puntsman constantly writes the name thus: "Yee-arle," an admirable example of phonetic orthography. Curiously, the Geese are never spoken of except in the plural. The word *Goose* is barred, and a single bird will be described as "a lame *Geese*"!

DIFFICULTIES AND DANGERS OF THE GUNNING-PUNT.

I HAVE already stated my experience that during mild winters, and under certain conditions of weather, wildfowl are practically unobtainable, even with the complete appliances of punt and stanchion-gun. I refer to the ordinary type of gun as generally used on the coast. Of course, if the size and weight of wildfowling weapons is to be increased indefinitely, the case might be wholly altered; but surely there should be some limit in this direction. Geese might, no doubt, be reached with *shrapnell* at many hundred yards, for a season or two; after that they would leave this country for good.

The unfavourable conditions referred to sometimes prevail throughout the whole winter season, and perhaps not a score of Geese will be killed by all the gunners of the coast, though hundreds, or even thousands may daily be seen. Yet there have been written books which profess to tell us how to get them—and to get them in numbers—under any condition whatever. Such conceits would, at any rate on our bleak and shelterless N.E. coast, prove quite misleading, and probably lead their would-be exponents into no small personal danger. A moderately, if not entirely calm sea, is an absolute essential in punt-gunning, and how seldom do we have it in winter? How often does a wildfowling diary contain such entries as "Strong breeze from E. all day; could not go afloat"; or else, "Blowing a whole gale this morning; harbour one sheet of white water." Day after day, perhaps a whole week, may be lost thus, and the fowler can only wait, smoke, watch the glass, or (best of all) go home.

A gunning-punt, to be of any real service, must necessarily be a very low and shallow craft. Her "depth of hold" is only some nine or ten inches, and her freeboard perhaps four or five. In such a vessel it is obviously the height of recklessness and folly to venture into rough water. The slightest sea is liable, and certain, to break on board, placing her crew in the utmost discomfort, to say nothing of danger. Even with the most cautious, however, it will sometimes occur that, from a sudden change of wind or other cause, a puntsman will be placed in a position, if not of actual danger, at least of very great discomfort and difficulty. As an example of this, the author well remembers one January day when, on the wind suddenly shifting from W. to N.E., with furious squalls and driving snow, he was placed on a lee shore, where the rotten mud prevented a landing, and with some six miles of rough water to face ere shelter could be hoped for. We had hardly time to unship the gun and bring her inboard, so as to trim the punt by the stern, ere the change was complete. It then only remained to sit low and pole like a bargee, using special care to keep the small craft head to sea. Once let her fall off, and the rush of a sea take her in flank, and it is all up :—

. . . . tum prora avertit : et undis
Dat latus : insequitur cumulo præruptus aquæ mons.

At intervals one of us had to take a turn with the bailing-scoop to keep us from swamping through the seas that broke inboard. In the centre of the slakes, where the water was deep, it sometimes looked as though we were not going to get through ; but at that point we luckily began to meet the drift ice which had been accumulated by the tide and W. wind along the eastern shores. This circumstance perhaps saved us, for though the ice added greatly to the labour, it had the effect of "flattening" the sea, which no longer broke into us ; and we eventually reached the shore safe, but not very comfortable.

This incident, however, is mere child's play as compared with the narrow escape of my brother Alfred in the Holy Island slakes on that fearfully memorable day, the 14th of October, 1881, a day which strewed the N.E. coast with the bodies of fishermen and the shattered wrecks of their vessels. The morning broke fine, but the barometer having rapidly dropped to 28° 40′, the Northumbrian men did not put to sea. Those of Eyemouth and Burnmouth, a few miles to the northward, set sail as usual, and encountered the full fury of the cyclone, with the melancholy result that 170 lives were lost, and the manhood of the latter village all but annihilated.

The following is my brother's description of that terrible day :—" Launched the punt at seven and poled up the North Slakes with the flood, picking up a couple of Godwits with the small gun. The day was fine, though rather windy, and the sky was full of white fleecy clouds. About eleven o'clock we made out a large flight of Wigeon, splashing and gambolling by the edge of a mud-bank. They were very restless, rising at 300 yards without any warning; they made several impetuous gyrations in the air, sometimes passing close over the punt, where we lay flattened on the bottom-boards; but so unsettled were they (probably conscious of the coming storm), that, though frequently splashing into the water almost within shot, they never allowed time to execute the necessary manœuvres for a shot with the big gun. At noon we anchored to have our lunch, before going into action again. The weather had not altered—fleecy clouds still hastening across the sky from the westward. But now a small black cloud seemed to rise from the northern horizon. Quickly it rose and increased in magnitude. Hardly had we observed its ill-omened appearance, than a horrible roaring noise was borne down upon us. Up went the Wigeon and away to the southward, followed by a string of Grey Geese, driving before the wind at enormous velocity. At these I sent a No. 1 wire-cartridge, and remember distinctly hearing it patter against their strong quills as they hastened before the cyclone; now a flock of Oyster-catchers, now a score of Godwits hurried past. The rushing noise increased in violence. We could now see its advance, as the once tranquil waters were lashed up into seething foam some 500 yards to windward. 'Be quick, sir! jump out, and hold on fast!' shouts S——; but it was too late. In an instant we were overwhelmed—the punt bottom upwards, myself struggling to get out from underneath her. Being a mile from the main shore, it was a case of hold on, or drown. Never have I seen such a fearful sight as those masses of seething waters. The wind roared like an express train bearing down on one in a subterranean tunnel, scooping up the water and flinging it in our faces. So terrific was its force in these wide and unprotected slakes, that it was impossible to stand or to make our voices heard to each other. As we were carried along in the driving foam, we simply clung to the coamings of the punt, and twice did the wind pick her off the water and literally hurl her over me, who happened to have the leeward berth, S. hanging on to windward. Fortunately, we managed to hit off a 'gut,' or stream, leading up to the mainland. The depth varied here from two to four feet, and in places the wind fairly scooped the water out of its channel, dashing it bodily upon the black ooze which formed its sides. This ooze was, if possible, more treacherous than the water, affording no foothold, and, on the contrary, tending to anchor one in its slimy depths. In the midst of the hurricane, I well remember seeing close at hand a belated Oyster-catcher thrown down on the mud, and how quickly the poor bird instantly headed to windward. Gulls, too, and the smaller waders seemed equally incapable of flight, as frequently they were dashed down on the muddy ooze, and, crouching low, tried to seek shelter there from the overwhelming elements.

"For three hours and a half we struggled thus with the storm, fighting foot by foot to gain the mainland, which lay on our beam, as we faced the gale. Eventually we reached the shore near an old barn. To this we crept on all fours, and while lying, half-unconscious, behind its gable, the slates were lifted high in the air and carried away. During these three and a half hours, some hundred and seventy fishermen were drowned within a few miles of us, and twice we observed the glare of the rockets as they conveyed the life-saving apparatus to the crews of two stranded steamers a couple of miles to seaward of us. That night I spent at the farm-house at Fenham, where the good-wife wrapped me in blankets and acted as a veritable good Samaritan. As a warning to brother sportsmen, I would urgently advise never to go out punting without having observed the barometer before starting. I afterwards found that the glass that morning had stood at 28° 40′ (sea level)—a sufficient warning of the impending hurricane."

THE LAST DAY OF WILDFOWLING.

A LUCKY WIND UP.

After a mild and open winter, which on our coast had been almost totally unproductive of fowl, we betook ourselves on the evening of the 27th February to the remote fishing hamlet at which we have long established our wildfowling headquarters to try our luck in a final day's campaign, for the end of the season was rapidly approaching. By the way, it may be remarked that there is no conceivable reason why Wild Geese and Wigeon should not be killed in March. The legal restrictions are a cruel injustice to many a poor fisherman-fowler; but figs do not grow on thorns, and it is about as reasonable to expect politicians to understand such subjects as to ask a punt-gunner to settle the Irish question. Well, it was dark enough as the slow train at last pulled up at the roadside station, but presently the clouds passed away, the full moon shone out, and the long radiating columns of the "northern lights" flickered brightly across the heavens as we traversed the water-logged sand-flats. The longest road has an end at last, and we are presently hard at work on the ham-and-eggs in our snug den, in a hamlet redolent of fish and things piscatorial. There only remained to us one day to shoot (for the 1st of March was a Sunday), and we determined to make the most of it, albeit our chances of success in so extremely mild a season were very remote. Accordingly, after a couple of hours' "snooze," on two hard oak chairs, I turned out at midnight, and passing through the tortuous little street, paved with the shells of defunct generations of mussels and cockles, proceeded to launch our trim little craft, the

A SCIENTIFIC SHOT: GEESE COMING OVER—HIGH.

To face page 276.

Boanerges gunning-punt. First the big gun had to be loaded. Down her long barrel rattle the 32 drams "Colonel Hawker," followed by 10 ounces of BB—the priming is carefully inserted, and the cap fixed. Then she is gently adjusted into position—the gear, ammunition, &c., all stowed, everything in its place, for aboard a gunning-punt there is not a square inch of room to spare—and away we go. With a brilliant moon, a dead-calm sea, and a flowing tide, we proceed right merrily, and our hopes rise rapidly—on so favourable a night surely we shall manage a heavy shot at the Wigeon! But it was not to be. At 3 A.M. a change came over the scene. The western horizon suddenly banked up with cloud masses, and we presently heard afar that strange rustling sound, like the distant rumble of an approaching express, which at sea foretells wind. On it came. In ten short minutes the driving clouds were scudding across the moon, and what had been calm white water was lashed into a confused black mass. For some time we persisted in shoving to windward, but our efforts to gain the weather-shore were in vain. Sea after sea broke into us, and the chance for the night was clearly gone. For a couple of weary hours we sought shelter on a desolate bent-grown sandspit. Then the ebb tide forced us to quit this refuge, and make the best we could of our passage back—to bed. Thus ended attempt No. 1—a failure; and No. 2 failed likewise. Before day-break we were out again at the "morning flight," but though we saw plenty of fowl, with a fair show of Geese, not a single shot rewarded our two hours' vigil, and at 9 A.M. we returned a second time empty-handed (and empty elsewhere) to breakfast. Such, in plain fact, is but the common luck of coast wildfowling—the most difficult, uncertain, yet withal one of the most exciting of all our British sports. It is amusing to read of hecatombs of the wary birds slaughtered on paper, and still more so to see the lightsome mood in which the undertaking is often essayed by "'prentice hands"; but after years of practice the writer can confidently state that, though patient, dogged perseverance and skill will from time to time reap their due reward in most gratifying success, yet there is no royal road thereto, nor on salt water will duck-shooting and shooting ducks ever become synonymous terms.

To return to our narrative. It is still only 10 A.M., and there remain to us several hours in which to avert the disaster of an empty bag. So we start on our third essay. The tide being now about dead low, the field of operations is restricted to the deep-water channels which intersect the mud-flats. These vast expanses of ooze—too soft to carry a man, too solid to float a punt—thus, through the action of the tides, afford a safe asylum to the fowl, where twice every day they can for several hours feed and rest in peace, secure from man and all his works. At first luck seemed about to dawn upon us, for in the channels we got a pair of Scaup-duck and a Grebe with the small gun. Then the current changed again. As the tide flowed over the mud, we observed that a number of Mallard had remained "inside" to feed on the luxuriant crop of sea-grass—*Zostera marina*—which now waved in the tide currents beneath us in luxuriant swathes of brightest emerald. At the Mallard we had two punt-shots during the flood: the first, at about a dozen sitting scattered on the mud-edge, was a total failure. The range was the deadly seventy yards, the elevation was correct; but, though the BB seemed fairly to rake their position, not a single bird stayed. It was one of the mischances that will occur; so with an effort we choke despair and try again. The other shot was at rather over a score swimming in roughish water and at a longer range. As the smoke cleared off we saw four Mallard stretched on the sea and two more fatally crippled: still far from being as satisfactory a shot as it might have been.

All the morning we had had the Geese in view, some busily feeding in black patches on the *Zostera*, others flying restlessly about in long gaggling skeins. The winter had been so mild and open throughout that these wary fowl were far too watchful to allow the slightest chance of approach, even in a punt. Now, at full tide, they lined the shore in scattered companies for several miles, and their white sterns were conspicuous bobbing up among the dark wavelets as they reached down to the succulent sea-grass beneath them.

From midday till dusk we stuck to them. Every "dodge" we knew was tried. We "set" to them, sailed, paddled, drifted—all in vain. Hour after hour slipped fruitlessly away, and the only result of our manœuvres was that towards dusk we had their scattered companies now all congregated into one solid, compact phalanx of Geese, perhaps a thousand strong. There they sat, only half a mile from us, and just as the sun "took the hill" we commenced our last supreme effort. Alas! it was now a full quarter's ebb, and before we had approached within two gunshots of them we took the ground. Just then S—— nudged me and pointed out a couple of score of "Whews" (Wigeon) sitting on the mud-edge away to the right of us. Half in despair we hove our bows round to starboard to give them a trial; but it, too, failed. As luck would have it, a single pair was swimming, unobserved, in the black water between us, and these, rising close at hand, shifted the rest.

There now only remained the Geese, far up on the slobby ooze. As a final tactic, I determined to try for a flying shot (by "tipping" the big gun) as they went to sea at night. Accordingly we let the punt drive with the tide till we lay directly on their course to the seaward channel. Then we shoved in as near the mud-edge as was safe to go on the ebb, and waited patiently. The night was still and calm, the western sky aglow with the glorious hues of sunset, and not a sound audible but the gentle lapping of the tide against the punt, and the loud and weird babble of voices from the thousand throats in front of us. What a concert! No music sweeter to my ear; no articulate words more expressive of intensely watchful security, of guarded suspicion, than their varied intonations. At last the critical moment arrived, the moment which was to decide all our hopes and fears. "They're up!" With a roar like the distant rumbling of thunder, the sonorous host take wing for the open sea. Will they come our way? . . . Yes! straight for us, prone on our chests, head the leaders, and in ten seconds the sky above us is flecked with moving masses, and seamed with strings of "black half-moons." "Now then, sir! let 'em have it!" hisses S——, as I fumble for the eighth part of a second with the trigger-string (for "tipping" a punt-gun is no child's play); then up goes the long barrel, and afar across the darkening waters resounds her thunderous boom. Ye gods! I'm among 'em! right in the thick of them! Mark! three—five—six—seven—eight—fall all round us; fall in curving lines, each with a sousing "flop" into the sea, while at least two more slant away, body-struck, to fall dead a little further out. It was a glorious shot for a tipped one. But there is no time to revel in the triumph of the moment, for only one of our Geese lies actually dead, and "clear the decks for the cripple-chase" is the order of the day. Then for a long half-hour we pole and shove, as no galley-slave ever toiled before; we toil till the perspiration half blinds us, banging away the while with the cripple-stopper till all our "pensioners" lie stretched and prostrate on the sea. Then, with joyous hearts and a full forepeak, we set our sprit-sail, out centre-boards, and spin away homeward with wind and tide at eight knots through the gloaming, delighted with the final success of our eighteen hours' toil, and its reward in *the best shot of the season.*

"A PENSIONER.'

INDEX.

Q.

R.

S.

T.

Woodfall & Kinder, Printers, 70 to 76, Long Acre, London, W.C.